ENVIRONMENTAL REMEDIATION TECHNOLOGIES,
REGULATIONS AND SAFETY

THE ACTIVATED
SLUDGE PROCESS

METHODS AND
RECENT DEVELOPMENTS

ENVIRONMENTAL REMEDIATION
TECHNOLOGIES, REGULATIONS
AND SAFETY

ENVIRONMENTAL REMEDIATION TECHNOLOGIES,
REGULATIONS AND SAFETY

THE ACTIVATED SLUDGE PROCESS

METHODS AND RECENT DEVELOPMENTS

BENJAMIN LEFÈBVRE

EDITOR

nova
science publishers
New York

NOTICE TO THE READER

Additional color graphics may be available in the e-book version of this book.

Library of Congress Cataloging-in-Publication Data

ISBN: 978-1-53615-202-9

Published by Nova Science Publishers, Inc. † *New York*

CONTENTS

PREFACE

In this book, the authors report on the pretreatment methods for waste activated sludge based on pulsed electric field and corona discharge techniques. The effects of pulse magnitude, frequency, temperature and pretreatment time are demonstrated on the basis of cell membrane electroporation. The influence of voltage polarity, frequency, magnitude, treating time and temperature has also been demonstrated.

A description of fundamental techniques in molecular biology for the analysis of the microbiota of activated sludge is provided. Activated sludge is a heterogeneous system of organisms, organic and inorganic material, and therefore giving a specific protocol for each molecular technique would be imprudent.

The authors go on to discuss the Monod model, which provides a functional relationship between specific growth rate and substrate concentration in the bulk. Important research efforts dedicated to adequate use of the Monod model are presented, consolidating knowledge from activated sludge and biofilm modelling, identifying misdirections, and setting parameters for further research.

In one study, different microwave power outputs and times were optimised for sludge solubilisation without evaporation loss in waste activated sludge from two different sources. The variable effects of pre-treatments on extracellular polymeric substances fraction, cellular oxidative stress and solubilisation of both sludges were evaluated to understand the impact of sludge complexity.

The penultimate chapter examines how toxic carbon sources can cause higher residual effluent dissolved organic carbon than easily biodegraded carbon sources in the activated sludge process. Based on the variations of chemical components of activated sludge, mainly intracellular storage materials, extracellular polymeric substances and soluble microbial products, the performance and mechanism of toxic carbon on the activated sludge process can be clarified.

The purpose of the final study is to research the supplementation of different concentrations of substrate on the degradation rate of xenobiotics, and to determine the optimal concentrations of auxiliary substrates that are most beneficial. The results show that sugar and peptone can affect 2,4-D degradation rate by several different degrees at different concentrations.

Chapter 1 - Waste activated sludge (WAS) is a main byproduct of wastewater treatment through an activated sludge process that could bring further pollution to the environment and requires disposal. Anaerobic digestion (AD) has been widely used to dispose of the WAS, however, the efficiency of AD is usually restricted by the sludge hydrolysis step as a lot of organic matters are enclosed in the microbial cells of sludge and can't be biodegraded directly. A step called sludge pretreatment should be performed before the AD to disrupt the microbial cell and release the intracellular organic matters for biodegradation. In this chapter, the authors report on pretreatment methods for WAS based on pulsed electric field (PEF) and corona discharge techniques. Three parts with repect to the pretreatment behavior are included. First, the pretreatment on WAS by PEF is stated. Effects of pulse magnitude, frequency, temperature and pretreatment time on the efficiency is demonstrated on the basis of cell membrane electroporation. Second, corona discharge based pretreatment is achieved. The discharge is triggered by DC and high frequency AC voltage. The influence of voltage polarity, frequency, magnitude, treating time and temperature on the efficiency has been demonstrated. Third, a combined pretreatment method based on DC corona assisted PEF technique will be presented. The dependence on the efficiency of DC voltage, polarity, and

PEF magnitude is discussed. Finally, conclusions are given to summarize the keypoints of WAS pretreatment by the PEF and the corona discharge techniques.

Chapter 2 - In this chapter, a description of the fundamental techniques in molecular biology for the analysis of the microbiota of activated sludge is given. Because activated sludge is a heterogeneous system of organisms, organic and inorganic material, givig a specific protocol for each of the molecular techniques would be imprudent. This is because even the same composition of the activated sludge generates a challenge in terms of proper handling and treatment of the samples. In addition, further barriers include the type of equipment available in the laboratory and economic resources. Microorganisms are fundamental in wastewater's biological treatment; the cleaning of water relies on the settleability of activated sludges. The term of activated sludge is commonly used to define a heterogeneous assembly of microorganisms organized in aggregates named "flocs." Aggregation refers mainly to bacterial production of exopolymers that promote attachment of microorganism and inert matter. However, filamentous organisms are an essential part of the floc population in an activated sludge process, forming the backbone to which floc-forming bacteria adhere. But, a high proliferation or lacking filamentous bacteria in the sludge can affect the operation conditions of a wastewater treatment plant due to the formation of pinpoint flocs, filamentous bulking, and viscous or zoogleal bulking. The pinpoint flocs are small and mechanically fragile, presenting low settleable properties due to the lack of a filamentous bacterial backbone. Filamentous bulking is generated by the filamentous bacteria overgrowth in the sludge, leading to poor sludge settleability, i.e., poor thickening characteristics of the sludge. Viscous bulking is caused by an excessive amount of extracellular polysaccharides (EPS), inducing a negative effect on the biomass thickening and compaction due to the water-retentive nature of EPS. It produces an activated sludge with a density closer to that of the surrounding water, increasing the sludge volume index, and, in some severe cases, there is no solids separation. In addition, some filamentous bacteria could induce foaming, which affects the operational conditions of the plant. These microbial phenomena can be summarized in three factors that impact the efficiency of the wastewater treatment process: filamentous bacteria, production of EPS, and floc-forming bacteria. The past decades have witnessed a significant growth in the use of optical and electron microscopy to analyze the development of activated sludges and identify filamentous bacteria. These techniques have increased the knowledge on the structural characterization of activated sludge, including morphological, physical, and chemical parameters that are closely related to solid–liquid separation. Image analysis methods hava been developed also to quantify the abundance of protruding filamentous bacteria based on filament length relative to floc area, but the activated sludge is not a mono-culture, a diversity of microorganisms exists with different morphological and physiological characteristics. Thus, molecular tools represent a big ally for further identifying filamentous bacteria, i.e., they allow determining phylogenetic microbial diversity in the activated sludge, identifying the microbial species and their abundance. Phylogenetic information can be the basis to determine the physiological status of activated sludge, recognizing which are metabolically active microbial populations, likewise, it allows monitoring of cultivation-independent phylogenetically-defined populations in the activated sludge, leading to a dynamic study of the wastewater depuration process.

Chapter 3 - The Monod model gives a functional relation between specific growth rate and substrate concentration in the bulk; the maximum specific growth rate and the half-saturation index being the parameters. State of the art activated sludge models are based on the Monod model and have found wide application in engineering practice. However, the Monod model remains empirical only and a fundamental explanation of the Monod model remains unknown, although it is believed that the values of the Monod parameters are a function of both diffusional mass transport outside and inside the activated sludge floc and enzyme kinetics inside the cells. An explanation of the Monod model might therefore be given by Fick's laws of diffusion and the law of mass action (or Michaelis-Menten kinetics). The application of Fick's laws of diffusion requires, however, spatial dimensionality, which is a problem since activated sludge flocs have a complex three-dimensional structure, a feature that seems to be elusive to be described adequately by a mathematical model. Note, that the same problem is faced when modelling biofilm systems, and therefore zero-dimensional biofilm models have recently been proposed, overcoming the problem of modelling the complex biofilm structure. The zero-dimensional modelling approach using the Monod model has thus been established in activated sludge modelling and analogously in biofilm modelling, bearing in mind that both systems consist of similar cell aggregates. The Monod model as used in the zero-dimensional approach describes macrokinetic behaviour of biological wastewater treatment systems rather than intrinsic kinetics of activated sludge flocs or biofilms. This comment will briefly review important research efforts dedicated to an adequate use of the Monod model, consolidate knowledge from activated sludge and biofilm modelling, identify misdirections, and set directions for further research towards a unified macrokinetic use of the Monod model in biological wastewater treatment.

Chapter 4 - Anaerobic digestion of waste activated sludge helps in sludge mass reduction, biogas production and sludge stabilization. This process involves a series of biological reactions viz; hydrolysis, acedogenesis, acetogenesis and methanogenesis that are carried out by respective microbial communities. The limiting factor for anaerobic digestion is slow hydrolysis rate. Various sludge pre-treatment methods such as advanced oxidation process (AOP), thermal, microwave, electrochemical, biological, ultra-sonication and hydrodynamic cavitation have been used to improve slow hydrolysis, dewaterability and reduce overall sludge volume. Pre-treatments have variable effects on sludge characteristics such as pH, chemical oxygen demand, total organic carbon, total protein and carbohydrate, zeta potential, suspended and volatile solid concentration. The pre-requisite of anaerobic digestion is to enhance solubility of particulate organic matter and facilitate digestion by hydrolytic bacterium. Conventional heat transfer in thermal treatment solubilizes insoluble organics and helps in odour control. Dielectric polarization caused by microwaves is responsible for solubilizing complex organic molecules, thereby effective in breaking EPS (Extracellular polymeric substances) and release of intracellular substances. Hydrogen peroxide (H_2O_2) is a strong oxidant, used in advanced oxidation process of waste water treatment. H_2O_2 undergoes Fenton type oxidation, leading to generation of hydroxyl radicals [OH^-] and superoxide [O_2^-] that oxidise various target compounds and causes cell lysis & EPS dissolution. Decomposition rate of H_2O_2 to [OH^-] is dependent on temperature. To accelerate [OH] generation, H_2O_2 is treated along with O_3, UV, ultrasound, thermal and microwave processes. In the current research work, different microwave power outputs (100W to 1100W) and time (1-3 min) were optimised for sludge solubilisation without evaporation loss in waste activated sludge from two different sources (NWAS & ATEAS). The variable effects of pre- treatments on EPS fraction, cellular oxidative stress and solubilisation of both sludges were evaluated to understand the impact of sludge complexity in pre- treatment. The optimum condition for maximum sludge solubilisation is 450W & 1% H_2O_2/TS and 880W & 1% H_2O_2/TS for NWAS and ATEAS, respectively. 30-40% higher sludge solubilisation (SCOD/TCOD) was observed in combined microwave and H_2O_2 oxidation treatment compared to individual treatment in NWAS. The combined treatment produced 8 and 4.1-fold higher s[OH^-] and s[O_2^-] respectively compared to the control experiment. It has been found that higher oxidative stress has corresponded to significant reduction in volatile suspended solids and chemical oxygen demand. Significant reduction in zeta potential in combined pre-treatment indicates an enhanced disaggregation and flocculation tendency in waste activated sludge. Thereby combined microwave and oxidation treatment improves dewaterability of sludge compared to individual treatments.

Chapter 5 - Activated sludge technology is the most used option for the treatment of wastewater. The toxic carbon source can cause higher residual effluent dissolved organic carbon than easily biodegraded carbon source in the activated sludge process. Based on the variations of the chemical components of activated sludge, mainly intracellular storage materials (X_{STO}), extracellular polymeric substances (EPS), and soluble microbial products (SMP), the performance and mechanism of toxic carbon on the reaction process of activated sludge can be clarified. In addition, the integrated activated sludge model based on carbon flows can be used to understand the mechanism. In the steady state, the toxic carbon can result in higher microbial cells death rate, decay rate coefficient of biomass, the utilization-associated products (UAP) and EPS formation coefficients, than that of easily biodegraded carbon, indicating that more carbon flows into the extracellular components, such as SMP, when degrading toxic organics. In the non-steady state, the yield coefficient for growth and maximum specific growth rate are very low in the acclimatization stage, while the decay rate coefficient of biomass and microbial cells death rate are relatively high.

Chapter 6 - The purpose of this study was to research about supplementation of different concentrations of the substrate on the degradation rate of xenobiotic and to determine the optimal concentrations of the auxiliary substrates that are most beneficial of xenobiotic degradation rate. 2,4-dichlorophenol acid (2,4-D) was used representative xenobiotic organic compounds, while peptone and sugar used for auxiliary substrates. The activated sludge was completely break down 100mg/l of 2,4-D for three consecutive times. The different concentrations between biogenic substracts of sucrose and peptone were fed separately or combined into the medium containing 200mg/l of 2,4-D and 140mg SS/l of activated sludge. The results showed that sugar and peptone could affect 2,4-D degradation rate to several different degree **at** different concentrations. In separate supplementation, 2,4-D degradation completed within 25 hours, 40mg/l sugar and 150mg/l peptone concentrations were found to be the optimal concentrations. In combined case, 2,4-D was consumed totally within 20 hours and the optimal concentration of the combined sugar and peptone concentrations were 40 **and** 150mg/l, respectively.

In: The Activated Sludge Process
Editor: Benjamin Lefèbvre

ISBN: 978-1-53615-202-9
© 2019 Nova Science Publishers, Inc.

Chapter 1

PRETREATMENT OF WASTE ACTIVATED SLUDGE BASED ON PULSED ELECTRIC FIELD AND CORONA DISCHARGE TECHNIQUES

Yu Gao[*], PhD, Yong Liu, PhD and Tao Han, PhD
School of Electrical and Information Engineering,
Tianjin University, Tianjin, China

ABSTRACT

Waste activated sludge (WAS) is a main byproduct of wastewater treatment through an activated sludge process that could bring further pollution to the environment and requires disposal. Anaerobic digestion (AD) has been widely used to dispose of the WAS, however, the efficiency of AD is usually restricted by the sludge hydrolysis step as a lot of organic matters are enclosed in the microbial cells of sludge and can't be biodegraded directly. A step called sludge pretreatment should be performed before the AD to disrupt the microbial cell and release the intracellular organic matters for biodegradation. In this chapter, we report

[*] Corresponding Author Email: hmgaoyu@tju.edu.cn.

on pretreatment methods for WAS based on pulsed electric field (PEF) and corona discharge techniques. Three parts with repect to the pretreatment behavior are included. First, the pretreatment on WAS by PEF is stated. Effects of pulse magnitude, frequency, temperature and pretreatment time on the efficiency is demonstrated on the basis of cell membrane electroporation. Second, corona discharge based pretreatment is achieved. The discharge is triggered by DC and high frequency AC voltage. The influence of voltage polarity, frequency, magnitude, treating time and temperature on the efficiency has been demonstrated. Third, a combined pretreatment method based on DC corona assisted PEF technique will be presented. The dependence on the efficiency of DC voltage, polarity, and PEF magnitude is discussed. Finally, conclusions are given to summarize the keypoints of WAS pretreatment by the PEF and the corona discharge techniques.

Keywords: waste activated sludge, anaerobic digestion, pretreatment, pulsed electric field, corona discharge

1. INTRODUCTION

With the rapid development of industry and urbaniztion processes, the production of municipal wastewater has increased for many years. Wastewater has become one of the most important environmental problems in the world [1]. Over the years, the activated sludge process has been utilized in a broad range of applications in wastewater treatment plants. A schematic diagram for typical steps involoved in such wastewater treatment processes is depicted in Figure 1 [2]. After the wastewater is disposed of by the activated sludge, a primary byproduct which is called the waste activated sludge (WAS) is produced. The WAS is a special substance that is featured with solid, semi-solid and liquid states. Figure 2 presents a photo of the WAS placed in a test tube and its microstructure obtained with a microscope. The appearance of the WAS is black, as shown in Figure 2(a), and it has a good viscocity. The component of the WAS is rather complicated. It contains a great deal of pathogenic microorganisms, parasites, unstable organic pollutants, heavy metal and other harmful substances [3]. If the WAS is released without necessary treatment, the toxic substance within it and the odor it generated would bring further environmental pollution. Moreover, the production of the sludge is high. As reported in literature, the amount of dry sludge produced in Canada and the European Union was 6.7×10^8 kg and 10^{10} kg respectively in 2005. The United States produced nearly 8.2×10^9 kg of dry sludge in 2010, and the production of dewatered sludge (with 80% water content) in China was 3.48×10^{10} kg in 2011 [3-5]. Accordingly, from the viewpoint of environmental protection, it is of great importance to dispose of the WAS.

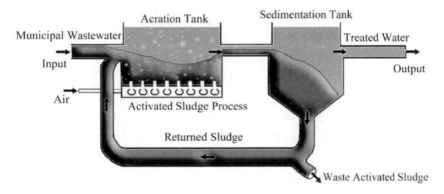

Figure 1. Typical steps of municipal wastewater treatment by activated sludge process.

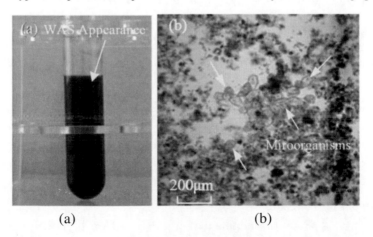

(a)   (b)

Figure 2. Typical characteristics of the WAS (a) overall appearance (b) microstructure.

Anaerobic digestion (AD) has been widely used for the WAS treatment which not only reduces the solid weight but also recovers energy in the form of CH_4 [6]. The AD is composed of three basic steps, i.e., sludge hydrolysis, fermentation of solubilized organic matters to form fatty acids and H_2, and methanogenesis of the fatty acids and H_2 [7]. Because the organic matters in the WAS are mainly presented within the microbial cells, they are unavailable for straight biodegradation that makes the sludge hydrolysis a rate-limiting step in the AD process [8]. The microbial cell membrane can act as a barrier so that the release of intracellular organic matters to the extracellular solution is limited [9]. In order to enhance the rate of sludge hydrolysis, a procedure called WAS pretreatment should be performed to disrupt the microbial cell, hence to release the organic matters for biodegradation [10]. Over the past 30 years, a number of pretreatment methods have been proposed based on thermal, chemical, mechanical and electrical mechanisms [11], and pretreatment technologies such as pyrolysis, ultrasonic, microwave, alkaline hydrolysis are put forward and their performances on disrupting the microbial cell membrane have been estimated. S. Sawayama et al. reported that the efficiency of AD could be improved by thermal pretreatment, which was ascribed to the degradation of the cell membrane [12]. V. Penaud et al. found that the alkali pretreatment could make the chemical oxygen demand (COD) 75%~80% with proper NaOH content when the sample was heated at 140°C for 30 min [13]. I. W. Nah et al. revealed that the mechanical pretreatment was able to damage the microbial cells to release intracellular organic matters, hence enhancing the solubilization of the WAS [14]. Although the successful applications of the above methods have been reported, the disadvantages of these technologies, like high energy consumption, low efficiency and complex operational procedures, have been realized as well. A novel pretreatment method with the characteristics of low energy consumption and easy operation is very urgent for WAS pretreatment in the future.

Recently, a WAS pretreatment method based on high voltage pulses has been developed [15]. As the high voltage pulses are applied on the pretreatment vessel, a pulsed electric field (PEF) will be established on the

WAS. The microstructure of the microbial cell in the WAS is changed when the transmembrane voltage (TMV) reaches several hundred mV [16]. Therefore, electroporation takes place on the membrane, which is featured with the appearance of numerous pores. The permeability of the membrane tends to be enhanced and becomes capable of releasing the organic matters [17]. Since the pulse is typically with a width of several tens of microseconds, the pretreatment is considered to be beneficial in energy consumption. Research works have been carried out to estimate the effects of PEF on WAS pretreatment efficiency. U. Koners et al. reported that the perforation of membranes resulted from PEF exhibiting less resistance against force which were able to release organic matters and to inactivate microorganisms in WAS [18]. J. S. Lee et al. found that the application of high voltage impulses could result in partial solubilization of the cell in WAS, and therefore the soluble chemical oxygen demand (SCOD) increased from 200 mg/L to 1700 mg/L with treating time of 90 min [19]. M. B. Salerno et al. revealed an increase of 60% in SCOD caused by the PEF application on the WAS, by which the sludge became more bioavailable for methanogenesis [20]. Il-Su Lee et al. reported on an increase in the hydrolysis rate of the WAS introduced by the PEF pretreatment. As a result, the promotion of 33% in methane production could be achieved [21]. B. E. Ritmann et al. compared the performance of methanogenesis based on the PEF and the ultrasonic methods and pointed out that the PEF pretreatment exhibited better efficiency [22]. Additionally, it was pointed out by H. Choi et al. that when PEF magnitude reached a value high enough, a discharge channel would appear within the sludge [15]. Although research has been conducted to exhibit the availability of the PEF application on the WAS pretreatment, the mechanism on how the PEF influences the disruption of microbial cells is still not fully understood.

On the other hand, corona discharge induced plasma has been widely considered as a promising method for sterilization in the field of medicine and fresh food keeping. The feasibility of its possible application on the WAS pretreatment has drawn much attention. M. A. Malik et al. claimed that the corona discharge was more effective, cheaper and eco-friendly as compared with the conventional water treatment methods [23]. The atmospheric corona discharge plasma generated by both DC and AC high voltage could evidently sterilize water from the majority of microorganisms [24]. V. Scholtz et al. compared the performance of microbial inactivation in liquid suspensions between the point-to-plane and the point-to-point electrode arrangements, and they found that the former electrode geometry led to higher treatment efficiency [25]. Z. Machala et al. confirmed that non-equilibrium plasma generated by three types of DC discharge (streamer corona, transient spark and glow discharge) in atmospheric air could be employed for the bio-decontamination of water [26]. I. B. Matveev et al. proposed that sewage sludge could be disposed to release or recycle with the application of thermal plasma [27]. Since corona discharge plasma exhibits a great potential in liquid organism inactivation, there also appears a possibility for corona discharge in the WAS pretreatment. The energetic particles, ultraviolet (UV) irradiation and chemically active radicals introduced by the corona discharge may sterilize the organism in the WAS, hence to release the intracellular organic matters.

In this chapter, the WAS pretreatment based on the PEF and the corona discharge has been introduced mainly on the basis of our research over the past 5 years. The possible mechanisms of the PEF and the discharge induced disruption of microbial cells in the WAS has been demonstrated. The influence of working parameters, such as PEF magnitude, frequency, temperature, treating time of corona discharge, on the WAS pretreatment performance has been analyzed. A PEF-corona discharge combined pretreatment method is proposed at last to take advantage of each technique.

2. EXPERIMENTAL SETUP

2.1. High Voltage Pulse Generator Design and Test Vessel for the PEF Pretreatment

In order to generate a PEF for the WAS pretreatment, a consecutive high voltage pulse generator has been desinged on the basis of the Marx generator principle. Figure 3 shows a schematic diagram of the electric circuit of the generator. The DC voltage is linearly controllable with a maximum output of 500 V, R_c is a protective resistor, $D_2 \sim D_6$ are rectifier diodes, $R_{i1} \sim R_{i6}$ are charging resistors whilst $R_{o1} \sim R_{o6}$ are discharging resistors, $C_1 \sim C_6$ are the capacitances for energy storage, $S_1 \sim S_6$ are insulated gate bipolar transistors (IGBT). The DC voltage is ultilized to charge the Marx generator, the IGBT is program controlled by a MSP430F149 singlechip, the output U_i from Ports a and b could be as high as 2 kV. As the Ports a and b are connected with the input Ports A and B of a pulse transformer, the output current from the Marx generator would act as the exciting current of the transfomer, by which a high voltage U_o could be formed at the output Ports A' and B.' The maximum output of the transformer is 65 kV, the rise time and the width of a single pulse are 5 μs and 40 ± 5 μs respectively. The maximum repetitive frequency of the pulses are 100 Hz. Figure 4 displays a typical waveform for the single pulse.

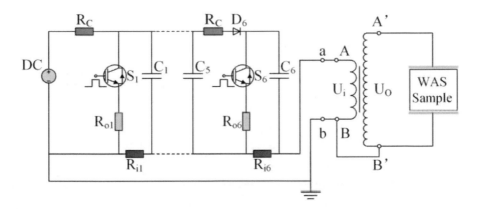

Figure 3. Schematic diagram of test circuit for high voltage pulse generator.

In order to pretreat the WAS with the high voltage pulses, a test vessel has been designed to generate a PEF. Such vessel is made from a petri dish which has an inner radius of 50 mm and a depth of 15 mm. As depicted in Figure 5, a pair of copper foils is attached through conductive adhesive on the top of the glass cover and the bottom of the petri dish to form a quasi-uniform electric field. The thickness of the foil, the cover and the petri dish are 100 µm, 1 mm and 1 mm, respectively. The copper foils are connected to the Ports A' and B' of the pulse transformer to generate a PEF on the WAS.

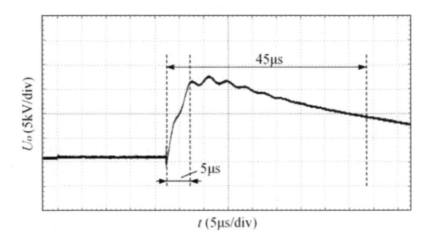

Figure 4. A typical waveform of the single pulse.

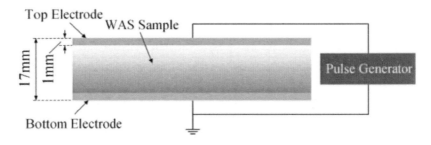

Figure 5. Test vessel for the WAS pretreatment with PEF.

2.2. Electrode Arrangement and Power Sources Used for Corona Discharge Pretreatment

Corona discharge is a special type of gaseous discharge occurring around the metal electrode with a small radius of curvature. Corona discharge triggered with different types of power sources would exhibit various features. In this chapter, the corona generated with DC voltage and high frequency sinusoidal AC voltage has been employed for the WAS pretreatment. The DC voltage is supplied by a commercial DC power source (DW-SA503-3ACF2, Dongwen Power Source Company, Tianjin, China) with maximum output of ±50 kV. The high frequency AC power source (CTP-2000K) is designed by the Corona Lab, Nanjing, China with the maximum output of 30 kV and frequency of 20 kHz. As the power sources are commercially available, the main work for the corona discharge pretreatment test is to design a suitable electrode configuration. In previous studies, several electrode systems for corona discharge generation have been proposed, among which the needle to plane and the needle to needle electrode geometry are usually employed. Considering the pretreatment efficiency, the discharge plasma formed with a single needle may be not sufficient. Therefore, a multi-needle to plane electrode system is designed for the study.

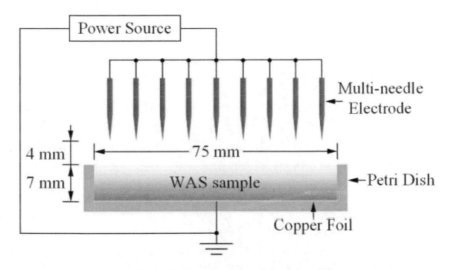

Figure 6. Schematic diagram of multi-needle to plane electrode for corona discharge.

Figure 6 shows the typical side view structure of the multi-needle to plane electrode system. The multi-needle electrode is composed of 61 needles. The plane electrode is formed by attaching a 100 μm thick copper foil onto the inner bottom of an acrylic petri dish. The petri dish has a diameter of 75 mm and a depth of 7 mm, allowing the WAS sample of 35 mL to be tested each time. The needle tip is positioned 4 mm above the WAS sample surface. During the test, the multi-needle array is connected with the power source, while the copper foil is grounded. Due to the small radius of curvature of the needle, corona discharge would be triggered around the needle tip. The discharge plasma as well as the oxidative species is then driven toward the WAS sample, by which the pretreatment of the WAS sample could be achieved.

2.3. Characterization of the WAS Pretreatment Performance

2.3.1. Coloring by Trypan Blue Solution

One of the most important purposes of the WAS pretreatment is to disrupt the microbial cell thus releasing the intracellular organic matters. Trypan blue solution has been employed to identify the disruption of the cell membrane. The molecular weight of a trypan blue molecule is ~960, which is not allowed to transport across the membrane. Once the cell is ruptured, the trypan blue molecule is capable of moving into the cell, and the cytoplasm of the cell is stained blue. On the contrary, the normal cell keeps its appearance. The trypan blue solution used for the identification is with a mass fraction of 0.4%, which is mixed with the sludge sample by a volume ratio of 1:9. Then the stained sample is placed under an optical microscope (XPS-3cb, Shanghai optical instrument factory, China) for visible inspection with a maximum amplification of 640.

2.3.2. Measurement of Released Organic Matter Content

As mentioned earlier, the SCOD is often employed to estimate the WAS pretreatment performance. SCOD is defined by the quantity of oxidant consumed by reacting with the reducing agent. As the soluble organic matter is the main reducer in the sludge solution, SCOD can be uscd as a measure of the organic matter in the WAS sample [19]. The measurement steps of SCOD are shown in Figure 7. A 6 mL sludge sample is injected into a 7 mL centrifuge tube, which is then treated in a centrifuge at a high speed of 13000 r/min for 30 min. 5 mL supernatant of the sample is picked out by a disposable burette and is filtered through four layers of slow filter paper to remove visible turbidity or impurity. The pore size of the paper is 1 μm ~3 μm. 1 mL filtrate is moved by a pipette to the digestion tube which contains 6 mL digestion solution of potassium dichromate. The digestion is performed at a temperature of 160℃ for 2 hours, after which the digestion tube is naturally cooled down to room temperature. Finally, 6 mL liquid within the tube is taken into the quartz colorimetric plate cuvettes and SCOD is measured with the spectrophotometric method. Furthermore, for the purpose of better exploring the pretreatment performance, ammonia-nitrogen (NH$_3$-N) as well as total phosphorus (T-P), which is also related to the content of intracellular organic matter, has been measured. The measurement steps of the NH$_3$-N and the T-P are similar to that for SCOD mentioned above except that the Nessler reagent and potassium persulfate are utilized as the digestion solutions, respectively.

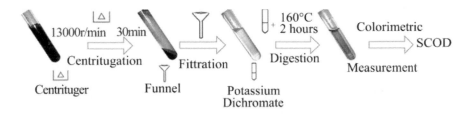

Figure 7. Measurement steps of SCOD for the WAS sample.

In this chapter, the WAS samples estimated are provided by a wastewater treatment plant (Research Institute Of Tianjin Capital Environment Protection Group Company Limited, Tianjin, China) at different time periods, thus the deviations of initial parameters of SCOD, NH$_3$-N and T-P are present. Accordingly, the relative change rates of SCOD, NH$_3$-N and T-P are employed to characterize the pretreatment efficiency of the WAS sample, which can be indicated as follows;

$$\Delta S\% = \frac{S_1 - S_0}{S_0} \times 100\% \tag{1}$$

$$\Delta N\% = \frac{N_1 - N_0}{N_0} \times 100\% \tag{2}$$

$$\Delta P\% = \frac{P_1 - P_0}{P_0} \times 100\% \tag{3}$$

where S_0 and S_1, N_0 and N_1, P_0 and P_1 are the SCOD, the NH_3-N and the T-P before and after the pretreatment. Another interesting parameter to indicate the degree of cell membrane disruption is the Intracellular Organic Matter Release Index (*IOMRI*), which can be expressed by;

$$IOMRI = \frac{S_1 - S_0}{S_T - S_0} \times 100\% \tag{4}$$

where S_T is the total SCOD of the WAS sample. The *IOMRI* is essentially a ratio of the released organic matter to the total organic matter in the WAS.

2.3.3. Other Characterization Methods for WAS Pretreatment

With the application of PEF and/or corona discharge, the membrane of the microbial cell in the WAS is possibly disrupted, hence the intracellular organic matters tend to be released into the sludge solution. Such behavior not only leads to the variation in the biochemical parameters as described in Section 2.3.2, but also results in the change in other properties like electrical property or pH value. The electrical conductivity and pH values are then measured and evaluated in the chapter. The electrical conductivity is measured through a conductivity meter (ECscan 30, Bante Instruments, Shanghai, China) with a measuring range of 0~20 mS/cm and an accuracy of 0.01 mS/cm. The pH value is tested by using a portable electronic pH meter with the accuracy of 0.01.

2.4. Test Procedure for the WAS Pretreatment

In order to evaluate the pretreatment performance of the PEF and the corona discharge method, three categories of experiments have been performed. In category I, the WAS pretreatment by the PEF has been carried out. The test is conducted by using the designed high voltage pulse generator and the test vessel shown in Figure 5. The test is undertaken within a temperature controllable thermotank with the maximum temperature range from 5^0C-90^0C and the accuracy is 0.1^0C. In this test, the temperature is set at 20^0C~60^0C, the magnitude of PEF is at 5.88 kV/cm~14.7 kV/cm, the pretreatment time is 10 min~25 min and the frequency of PEF is from 1 Hz to 100 Hz.

In category II, the WAS pretreatment is attained by employing the corona discharge method with the electrode configuration shown in Figure 6. Both DC and high frequency AC corona discharges are generated and their performance is estimated and compared. As regards the DC corona discharge, the test voltage is set at ± 1 kV~ ± 5 kV, the treating time is from 5 min to 60 min and the treating temperature is from 20^0C to 50^0C. For the high frequency AC corona discharge, the magnitude of AC voltage is set at 1 kV~5 kV as well. The frequency is controlled in the range from 5 kHz to 10 kHz. The treating time is 5 min and the temperature is 20^0C.

In category III, the WAS pretreatment is achieved by a PEF and DC corona combined method, which means that the WAS is disposed by the PEF after the DC corona. In such a case, the treatment protocol is scheduled as follows: the corona discharge triggered with DC voltage of ± 1 kV~ ± 5 kV is firstly applied to the WAS for 5 min at 20^0C, then the WAS is moved into a temperature controllable chamber to subject the PEF treatment. The PEF treatment is lasting for 10 min under 10 kV/cm to 30 kV/cm, and the temperature is kept at 40^0C.

Since the entire experimental work lasts for almost three years, the WAS samples provided at different periods of time by the wastewater plant vary in their initial parameters. For the simplicity of discussion, the WAS samples are referred to as Sludge Sample (SS) I and II. Some initial parameters for these samples are as follows. For SS I, the water content is

96%~97%, the total chemical oxygen demand (TCOD) is ~3500 mg/L, the initial SCOD is 150 mg/L to 450 mg/L. For SS II, the parameters are listed below in Table 1. All of the sludge samples are kept at 4^0C before the pretreatment tests.

Table 1. Initial Parameters for SS II

Item	Value	Unit
Water Content	95.91	%
Organic Content	50.96	%
TCOD	3500	mg/L
SCOD	300($\pm$150)	mg/L
Conductivity	2.5($\pm$0.2)	mS/cm
pH	6.82($\pm$0.1)	
NH$_3$-N	45($\pm$11)	mg/L
T-P	28($\pm$5.5)	mg/L

3. TEST RESULTS AND DISCUSSION

3.1. Pretreatment of the WAS with the PEF

The WAS as a complicated substance with solid, semi-solid and liquid states, contains a large number of organic fragments, microorganisms, inorganic particles and is anisotropic in nature. The type and the content of microorganisms are very rich. C. R. Curds et al. have studied the type of protists in sludge, it is found that flagellates, infusorians and sarcodina are rich in sludge [28]. M. Wagner et al. have investigated the type of bacteria in sludge, a number of proteobacteria and bacteroidetes are discovered [29]. The research results from M. Henze et al. indicate that the concentration of protists in sludge could reach 10^2~10^3/mL, while such a value increases to 10^7-10^9/mL for bacteria [30]. As microorganisms are exposed to the PEF, the arrangement of molecules on the cell membrane appears to be impacted by the field. When the PEF is applied on the WAS sample, the change in microstructure of cell membrane would take place on protist, fungus and bacteria within the sludge. A transmembrane potential (TMP) could be formed under the PEF to induce rearrangement in partial structure of the lipid bilayer, which results in the enhancement of permeability and conductivity of the cell membrane hence the "electroporation" phenomenon occurs [31]. Theoretically speaking, the pulse width should be larger than the charging time and the recovery time of the cell membrane such that the electroporation could be maintained. It has been revealed by J. C. Weaver et al. that temporary pores would be presented on the cell membrane as the cell is subjected to PEF with a magnitude of 1 kV/cm and width of μs~ms. Once the magnitude of PEF is high enough, the TMP is over the breakdown strength of the membrane and the irreversible electroporation could occur with the cell apoptosis [32]. In our research, the magnitude of PEF is in the range of 5.88 kV/cm~14.7 kV/cm and the pulse width is 35 μs~45 μs, which is sufficient to introduce the electroporation on the cell membrane. The appearance of arcella (typically with the size of 50 μm~100 μm, widely appears in the WAS sample), colored with the trypan blue solution before and after the PEF treatment is shown in Figure 8. The treatment is performed with PEF magnitude of 14.7 kV/cm, pulse frequency of 13 Hz, temperature of 40°C and treating time of 10 min. The sludge sample used in this section is SS I. For the arcella before PEF treatment as shown in Figure 8(a), its appearance is clean and transparent, which means that the cell membrane is undamaged, hence the trypan blue molecules can't get in. For the arcella exposed to PEF for a certain time, the color inside the cell becomes blue, which certainly indicates that the trypan blue molecules have been transported into the cell from the sludge solution. This should be recognized as evidence for the occurrence of electroporation. With the enhancement of membrane permeability, water molecules tend to move into the cell and to expand the cell membrane, as a result of which, the intracellular substances such as protein, lipid and carbohydrate are capable of releasing out to the sludge solution as well, therefore, the content of organic matters in the sludge will be enhanced. The pretreatment performance reflected by the disruption of the microbial cell can be estimated by measuring such content of organic matters.

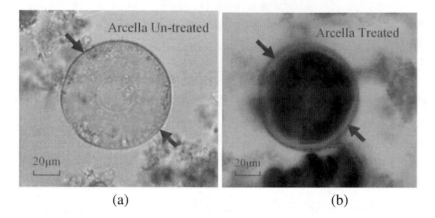

(a)   (b)

Figure 8. Typical appearance of microorganism in WAS sample colored with trypan blue before and after the PEF treatment. (a) Before treatment, (b) after treatment.

The typical relationship between the *IOMRI* and the electric field is shown in Figure 9. With the growth of the electric field, the *IOMRI* exhibits a non-linear increase. As the PEF magnitude *E* increases from 5.88 kV/cm to 8.82 kV/cm, a slight increase in *IOMRI* can be observed from 0.23% to 1.2%. However, a further increase of *E* from 8.82 kV/cm to 14.7 kV/cm leads to the enhancement of *IOMRI* to 9.07%. It is also found that the experimental results fit well with an exponential function as follows;

$$IOMRI = A \times exp(B \times E) \tag{5}$$

where *A* and *B* are the fitting parameters, depending upon the WAS sample and the PEF treating conditions. In this case, *A = 0.05* and *B = 0.35*. Although such an exponential manner is empirical, it can be understood from the viewpoint of biophysical nature as well. It has been proposed that the TMP is closely related to the magnitude of PEF. T. Kotnik et al. point out that for a monoplast with any shape, the TMP U_k generated under a bi-exponential electric field can be expressed by [16];

$$U_k = \alpha R_{cell} E \cos\beta \left(1 - exp\left(-t/\tau_m\right)\right) \tag{6}$$

where α is the shape parameter, for a circular cell, $\alpha = 1.5$; R_{cell} is the radius of cell; β is the angle between E and the normal direction of the cell membrane where the TMP is generated; τ_m is the charging constant of the cell membrane. It is then deduced from Equation (6) that with the increase of E or R_{cell}, the U_k appears to be enhanced. As a matter of fact, the radius of microbial cells within the WAS sample varies evidently. For instance, the radius of bacteria and fungus is about several to several tens of μm, and the radius of protist is in the range of several to hundreds of μm [33]. As the PEF magnitude is low, the protist like arcella with a larger size would be subjected to higher U_k, by which the electroporation appears easily for such kind of microbial cell. With the increase of the PEF magnitude, the U_k for the bacteria and the fungus with smaller size is going to be higher, hence the release of organic matters becomes more remarkable. In addition, it has been reported by A. O. Bilska et al. that with the growth of PEF magnitude, the electroporation becomes encouraged in such a way that the ratio of the electroporation area to the entire area of cell membrane increases non-linearly [34]. Accordingly, the exponential relationship between the *IOMRI* and the E could be understood from both the change in TMP and the variation in electroporation area in response to the PEF magnitude.

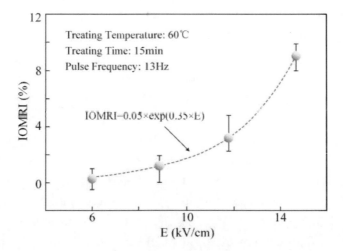

Figure 9. Typical dependence of the IOMRI upon the PEF magnitude.

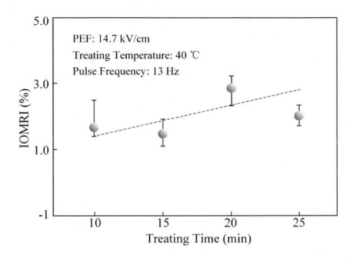

Figure 10. Typical dependence of the *IOMRI* upon the treating time.

The performance of PEF pretreatment on the WAS sample is dependent upon the treating time as well. Figure 10 depicts the typical relationship between the *IOMRI* and the treating time for the WAS sample. It can be seen that when the treating time lapses from 10 min to 20 min, the *IOMRI* exhibits a rather slight increase. Previous research has indicated that the electroporation under PEF can happen within a very short period of time. W. Frey et al. have studied the Jurkat cell colored with ANNINE-6 and have found that the TMP of the cell sharply increases in 5 ns with the application of PEF [35]. S. Talele et al. have reported that the TMP grows to its maximum value within 0.5 μs~1 μs under DC field and reaches stable status at 2 μs~3 μs [36]. Through a numerical calculation on lipid molecules with coarse-grain model, Q. Hu et al. have pointed out that when the TMP of a cell membrane reaches the threshold value, the electroporation will be formed within 5 ns~6 ns and gradually approaches a stable status [37]. In our test, the pulse has a rise time of 5 μs and a width of 45 μs. The TMP is sufficient for electroporation even with the application of a single pulse, leading to the release of intracellular organic matters and the increase of the *IOMRI*. With the extension of the PEF treating time, the occurrence period of electroporation that improves the permeability of the cell membrane tends to increase, thus the *IOMRI* is enhanced. Since the electroporation could be achieved within even one pulse, the increase in pulse number (the treating time) would not significantly encourage the further formation of pores on the membrane. Accordingly, only slight improvement on the *IOMRI* is observed with the treating time.

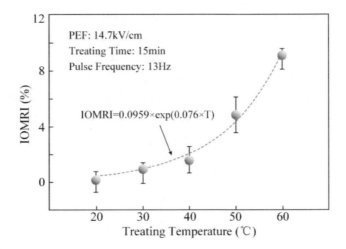

Figure 11. Typical dependence of the *IOMRI* upon the treating temperature.

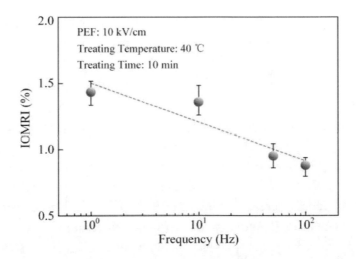

Figure 12. Typical behavior of the IOMRI as a function of PEF frequency.

However, the influence of the treating temperature on the *IOMRI* is very significant, as can be found in Figure 11. With the increase of the treating temperature, the *IOMRI* exhibits a remarkable increase that follows well with an exponential function. As the temperature is lower than 40^0C, the increment in the *IOMRI* with temperature is very small, which means that the effect of PEF is not so effective on disrupting the microbial cell in this temperature range. However, as the temperature is higher than 40^0C, the growth in the *IOMRI* with temperature is quite significant, indicating that the PEF would play a more important role under higher temperature conditions. The possible mechanism for such a behavior is that the cell membrane is so sensitive to the temperature. D. Exerova et al. have investigated the influence of temperature on the cell membrane. A phase transition at the membrane is observed when the temperature rises from 20^0C to 55^0C [38], which results in variations in the specific heat capacity and elasticity of the membrane [39]. At the region of the membrane where the phase transition occurs, the movement of membrane molecules is drastic. Such a movement in the molecular structure would have a significant impact on the PEF induced electroporation such that the pores are more easily generated. When the temperature is high, both the number and the size of pores are large [40] to allow more water molecules to get inside, and thereby the release of intracellular organic matter is encouraged. Furthermore, the diffusion of the organic matters released becomes accelerated at high temperature, which is helpful for further release of the matter. As a result, the *IOMRI* increases significantly at the higher temperature region.

The typical influence of the PEF frequency on the *IOMRI* is illustrated in Figure 12. The PEF treatment with a magnitude of 10 kV/cm, temperature of 40^0C and treating time of 10 min is taken as an example. It is surprising to find that the *IOMRI* decreases slightly from 1.43% to 0.87% when the frequency increases from 1 Hz to 100 Hz, which indicates that the performance of PEF pretreatment becomes worse with the frequency in such a range. In order to interpret such an anomalous behavior, a schematic diagram has been envisaged to reveal the possible dynamics in the release process of intracellular organic matter as shown in Figure 13.

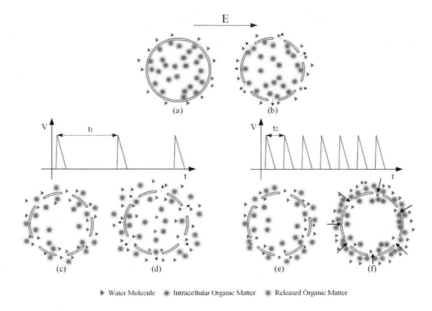

Figure 13. A schematic diagram of the release of intracellular organic matters under low and high PEF frequency. (a) the cell before electroporation, (b) the onset of electroporation, (c) and (e) the exchange of organic matter and water molecules under low and high frequency conditions, (d) and (f) the release and the accumulation of organic matters under low and high frequency conditions.

As can be seen in Figure 13(a), the cell membrane is undamaged such that the intracellular organic matter can't be released. With the application of PEF of a magnitude high enough, the electroporation phenomenon will occur on the membrane, hence some pores are formed to allow the exchange of water molecules outside the cell and the organic matter inside the cell, as shown in Figure 13(b). The released organic matter is initially accumulated around the pores, thereby restricting the further release of the matter. When the PEF frequency is low, the time interval t_1 between the consecutive pulses is comparatively longer. This provides more time for the accumulated matter to diffuse away into the sludge. Therefore, the concentration of matter around the pores is reduced and further release of matter is encouraged, as depicted in Figures 13(c) and 13(d). When the PEF frequency is high, the time interval t_2 between two pulses is short and the released organic matter doesn't have enough time to move away such that the further release is limited, as shown in Figures 13(e) and 13(f).

Consequently, PEF with high frequency leads to a slight reduction in the *IOMRI* as compared with that obtained with low frequency PEF pretreatment.

In summary, this section summarizes the effects of PEF magnitude, frequency, treating temperature and treating time on the pretreatment performance of the WAS sample by estimating the *IOMRI*. It can be concluded that the magnitude, the temperature and the time have a positive influence on the pretreatment, where the enhancement in the magnitude and the temperature would result in significant improvement of the pretreatment performance. The treating time has a slight influence on the performance as the PEF induced electroporation is considered to be achieved within a very short time. On the contrary, the frequency has a negative effect on the performance such that the increase in the frequency leads to a slight decrease in the release of organic matter, which is thought to be caused by the diffusion restriction of organic matter accumulated around the pores on the cell membrane. The results presented in this section would provide helpful information of the WAS pretreatment on the basis of PEF technique.

3.2. Pretreatment of the WAS with the Corona Discharge Method

Corona discharge is a special type of gaseous discharge which is usually introduced at the electrode with a small radius of curvature where the electric field distortion is strong. For the corona discharge occurring in atmospheric air, several physical and chemical processes would be involved, thus generating a series of oxidative species, energetic particles and UV irradiation [41]. Those associated effects with the discharge process have the ability to disrupt the microbial cell in the WAS sample, and the pretreatment by corona discharge may be achieved with proper type and parameters.

In the first part of this section, both DC and high frequency AC voltage are employed to investigate the corona discharge effect on pretreatment performance of WAS using SS I. Afterwards, in the second part of this section, the pretreatment was achieved by DC voltage using SS II to examine how the treating time and the temperature affect the pretreatment behavior.

3.2.1. Comparison of WAS Pretreatment Performance between DC and High Frequency AC Voltages Induced Corona Discharge

With the application of DC or high frequency AC voltage, the corona discharge is triggered at the needle tip in air, by which the WAS could be pretreated so that the disruption of the microbial cell and the release of intracellular organic matter are achieved. Figure 14 shows the appearance of a microbial cell (arcella as an example) before and after the corona discharge pretreatment. The cell before the treatment is presented in Figure 14(a). It can be observed that the cell structure is in good condition where no cracks or holes can be found at the membrane. However, as the WAS is exposed to corona discharge for a while, the arcella in the WAS becomes remarkably disrupted. Figures 14(b) and 14(c) depict the arcella appearances after corona discharge pretreatment by DC and high frequency AC voltage, respectively. The treating parameters are 1 kV, 5 min for DC whilst 3 kV, 10 kHz, 5 min for AC. It is observed that the cell is damaged and only the remains of arcella can be found within the WAS sample. The results shown in Figure 14 reveal the evidence that both DC and high frequency AC corona discharges have the ability to disrupt the microbial cell within the WAS sample.

The influence of the DC voltage on the $\Delta S\%$ for corona discharge pretreatment with both positive and negative polarities is shown in Figure 15. It can be seen that with the application of DC corona discharge, the SCOD becomes increased as compared with that before the pretreatment, which suggests that the DC corona pretreatment has the ability to damage the microbial cell and to make the WAS more bioavailable. The variation in the $\Delta S\%$ caused by the DC voltage is affected by voltage polarity. With the increase of positive voltage from 1 kV to 5 kV, the $\Delta S\%$ decreases from 15.6% to 6.8%. But with the negative voltage growing from -1 kV to -5 kV, the $\Delta S\%$ increases from 0.1% to 21.4%. It indicates that at lower voltage, positive corona discharge has better pretreatment performance than negative corona discharge, whereas at higher voltage the opposite tendency is true.

In order to better understand the effect of corona discharge on the pretreatment performance of the WAS sample, the corona discharge induced physical and chemical dynamics in the air should be analyzed. Corona discharge occurring in room air can generate a large number of energetic particles, UV irradiation and oxidative species. The electrochemical reactions between the energetic particles and the water molecule can be assumed as follows [42];

$$e^- + O_2 \longrightarrow O(^3P) + O(^1D) + e^- \tag{7}$$

$$O + O_2 + M \longrightarrow O_3 + M \tag{8}$$

$$O(^1D) + H_2O \longrightarrow {}^\bullet OH + {}^\bullet OH \tag{9}$$

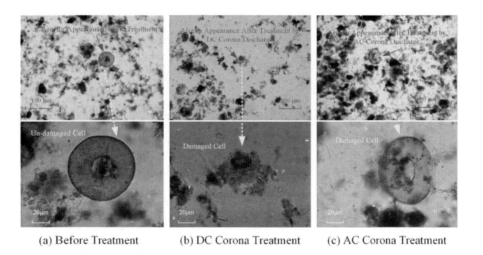

(a) Before Treatment   (b) DC Corona Treatment   (c) AC Corona Treatment

Figure 14. Appearance of a microbial cell before and after the corona discharge pretreatment.

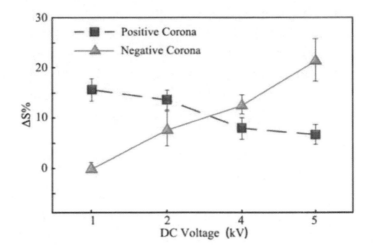

Figure 15. Effect of the DC voltage on the change of SCOD for both positive and negative polarities.

Among these reactions, ozone can transfer free electrons and H$^\bullet$ into hydroxyl (OH) radical [43]. With the formation of H$^+$ in positive corona discharge [44], the reactions can be described as follows;

$$H_2O \xrightarrow{e} H^\bullet + {}^\bullet OH \tag{10}$$

$$2H_2O \xrightarrow{e} H_2O_2 + H_2 \tag{11}$$

$$e^- + O_3 \longrightarrow O_3^{\bullet-} \tag{12}$$

$$H^\bullet + O_3 \longrightarrow HO_3^\bullet \tag{13}$$

$$O_3^{\bullet-} + H^+ \longrightarrow HO_3^\bullet \tag{14}$$

$$HO_3^\bullet \longrightarrow {}^\bullet OH + O_2 \tag{15}$$

The products with high oxidative nature, such as OH radicals, ozone, atomic oxygen and metastable oxygen molecules, are generated in the discharge process. Those oxidative species would induce chemical reaction with the cell membrane when they arrive at the WAS sample, which leads to the variation in cell microstructure. The cell membrane is made of lipid bilayers which contain a lot of unsaturated fatty acids [45]. A gel-like feature can be found at the membrane due to the presence of the unsaturated fatty acids, which encourages the transport of the biochemical byproducts across the membrane. The ozone and OH radical with oxidative ability will damage the lipids that act as a barrier against the transport of ions and polar compounds [46]. Another important component of the cell membrane is protein molecules. They are made of linear chains of amino acids, which are susceptible to oxidation by the atomic oxygen or the metastable oxygen molecules [47]. As stated above, the microstructure of the microbial cells within the WAS sample appears to be disrupted by the oxidative species generated by corona discharge. Furthermore, the consumption of OH radicals due to reaction with the membrane would encourage the decomposition of hydrogen peroxide [48], by which more OH radicals are generated.

$$H_2O_2 \xrightarrow{hv} 2\,^{\bullet}OH \tag{16}$$

$$H^{\bullet} + H_2O_2 \longrightarrow H_2O + {}^{\bullet}OH \tag{17}$$

$$e^- + H_2O_2 \longrightarrow {}^{\bullet}OH + OH^- \tag{18}$$

Accordingly, the disruption of the cell membrane is facilitated and the intracellular organic matters are released into the sludge solution. Moreover, the undissolved organic matter in the WAS sample could be oxidized to soluble matter as well, which contributes to the growth in SCOD.

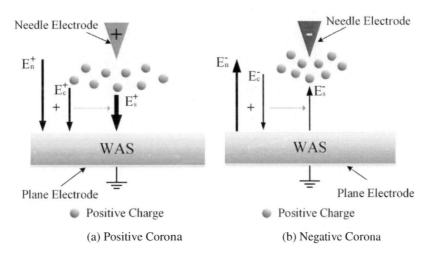

Needle Electrode

E_n^+

E_c^+

$+$ E_s^+

WAS

Plane Electrode

Positive Charge

(a) Positive Corona

Needle Electrode

E_n^-

E_c^-

E_s^-

$+$

WAS

Plane Electrode

Positive Charge

(b) Negative Corona

Figure 16. Schematic diagram of charge migration and electric field distribution in DC corona with different polarities.

The polarity dependence of DC corona pretreatment performance should be ascribed to the fact that positive and negative corona discharges provide different features of charge transportation. When positive voltage is applied between the multi-needle to plane electrode, electric field E_n^+ will be formed as depicted in Figure 16(a). With the occurrence of corona discharge around the needle tip, electrons run away via the needle whilst positive charges are residual in the vicinity of the needle electrode. Accordingly, the field E_s^+ above the WAS sample is superposed by E_n^+ and E_c^+ that is triggered by the positive charges, hence becomes enhanced, as shown in Figure 16(a). With respect to the negative corona discharge, positive charges are restricted around the needle tip whereas electrons drifted toward the WAS sample. An electric field E_c is established between the positive charges and the plane electrode, which is in the opposite direction of E_n, as shown in Figure 16(b). Accordingly, the field E_s above the WAS sample is weakened due to the superposition of E_c and E_n. A numerical simulation of electric field distribution above the WAS sample surface is also in support of such a demonstration, the simulation result is illustrated in Figure 17. Here the DC voltage used is ±5 kV, and the radius of curvature of needle tip is 30 μm [49]. It can be seen from Figure 17 that the gaseous ionization region Z1 is very small (0.27 mm), the electric field generated with positive corona is higher than that with negative corona in the charge transportation region Z2. Therefore, with the application of positive voltage, the energetic particles as well as the oxidative species are more easily transferred to the WAS sample, and the disruption of the cell membrane should be encouraged in such a case.

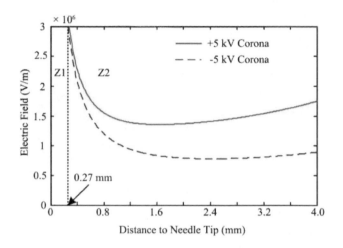

Figure 17. Electric field distribution between the needle tip and the WAS surface with the application of ±5 kV DC voltage [49].

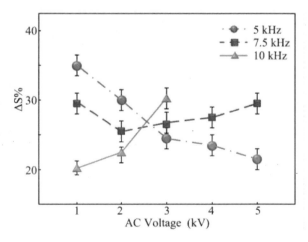

Figure 18. The influence of AC magnitude on the ΔS% with various frequencies.

However, as the positive voltage increases, the production of the oxidative species that are available for the WAS pretreatment appears to be enhanced. The species will then not only disrupt the microbial cell or degrade the undissolved organic matter within the WAS to result in the enhancement of SCOD, but also further oxidize the soluble organic matter into the form of CO_2 and H_2O which tends to reduce the SCOD [50]. Such an effect can be referred to as the "over oxidation effect" [51]. Therefore, the $\Delta S\%$ appears to decrease with the positive voltage. On the other hand, as the negative voltage increases, although the corona discharge is strengthened to generate more oxidative species, only part of them are accessible for the WAS since the main species are generated around the needle tip. Furthermore, the OH radicals could be consumed by CO_3^- produced in the negative corona discharge in the following way [51];

$$CO_3^- + \cdot OH + M \rightarrow M + HCO_4^- \tag{19}$$

This means that the OH radials in the negative corona discharge that is useful for the membrane disruption is not as sufficient as that in the positive corona discharge.

Therefore, the over oxidation effect mentioned above would not be present in the sludge treated by negative corona, and the $\Delta S\%$ appears to increase with the negative voltage.

When the WAS sample is exposed to the high frequency AC voltage with 5 kHz, 7.5 kHz and 10 kHz, the disruption of the microbial cell is observed as well. Figure 18 exhibits the change in $\Delta S\%$ caused by the increase of AC magnitude. Basically, the SCOD increases after the AC corona pretreatment, which confirms the availability of such method in the WAS pretreatment. The influence of AC magnitude on the $\Delta S\%$ has a strong dependence upon the AC frequency. When the frequency is 5 kHz, the $\Delta S\%$ decreases with the magnitude. When the frequency is 7.5 kHz, the $\Delta S\%$ initially decreases then appears to increase with the magnitude. At the frequency of 10 kHz, the $\Delta S\%$ grows significantly with the magnitude. However, as the AC magnitude is over 3 kV, a gaseous discharge occurs between the needle electrode and the WAS sample. Since the discharge nature is changed from the corona discharge with lower energy density to the spark discharge with higher energy density, the SCOD at 10 kHz with the AC magnitude higher than 3 kV is not measured. As compared with the results shown in Figure 15, it can be found that the pretreatment performance of the AC corona is better than that of the DC corona.

In order to understand the frequency dependence of the pretreatment performance, the possible mechanism with respect to the gaseous discharge under AC voltage is analyzed. When the sinusoidal AC voltage is applied between the multi-needle to the plane electrodes, the charges generated from gaseous ionization will be distributed in the air gap above the sludge sample which leads to field distortion, hence influencing further corona discharge. The electric field distribution and the charge migration are shown in Figure 19 as a schematic diagram. When the air gap between the needle tip and the sludge surface is exposed to the positive half cycle of AC voltage, the electric field above sludge surface E_s would be strengthened due to the superposition of the field generated by the external voltage and the positive space charge. Such a case is helpful for the production of oxidative species, and is encouraging for the disruption of microbial cells in the WAS. While the negative half cycle is applied, the positive charges are driven to the vicinity of the needle electrode, the field around the needle is enhanced and the gaseous ionization is encouraged. However, the field just above the sludge surface is suppressed to limit the transportation of oxidative species and energetic particles. In addition, the CO_3^- generated in the negative half cycle can consume the OH radicals, by which the over oxidation effect mentioned above can be inhibited. While the positive half cycle appears again, the positive charge would assist the improvement on E_s again that leads to the disruption of the microbial cell. Accordingly, it can be deduced that the positive half cycle may generally have a "negative" effect with the presence of "over oxidation effect" which leads to the reduction in SCOD, whereas the negative half cycle may have a "positive" effect as it balances the over production of oxidative species thus to enhance the SCOD.

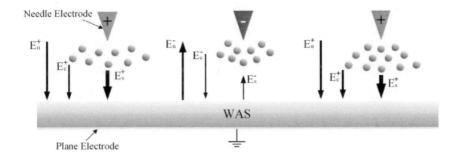

Figure 19. Schematic diagram of charge migration and electric field distribution under AC voltage.

During the application of high frequency AC voltage, the "negative" and the "positive" effects take place simultaneously, and the pretreatment result is determined by the one most dominant. As the frequency is 5 kHz, the time period at positive/negative cycle is relatively longer, the positive charge can migrate to a position far away from the needle, as a result of which the field E_s will be remarkably improved, leading to the higher production of oxidative species. Therefore, the "over oxidation effect" in positive half cycle would play a dominant role. With the increase of AC magnitude from 1 kV to 5 kV, the position of positive charge becomes closer to the sludge surface, thus the "over oxidation effect" is encouraged to decrease the $\Delta S\%$. With respect to the frequency of 10 kHz, the time period of a cycle is shortened, the positive charges are located closer to the needle. Accordingly, it limits the possible improvement of E_s on positive half cycle but encourages the gaseous ionization around the needle tip on negative half cycle, which makes the negative half cycle play a more important role in the WAS pretreatment. As a result, the $\Delta S\%$ increases with the AC magnitude. When the frequency is 7.5 kHz, the effect of AC corona discharge from positive and negative half cycles has to face a competition depending upon the AC magnitude. When the magnitude is lower than 2 kV, as depicted in Figure 18, the "over oxidation effect" that occurred on the positive half cycle is thought to play a dominant role, leading to the reduction in $\Delta S\%$. When the magnitude is higher than 2 kV, the negative half cycle becomes dominant, hence the $\Delta S\%$ appears to increase.

Although the high frequency AC corona discharge possesses a better performance on the disruption of microbial cells than the DC corona discharge, the energy consumption of the AC case is also higher than that of the DC case. In this chapter, the energy consumption ratio of the WAS pretreatment η is defined as the energy consumed by the sludge of unit volume with the increase in $\Delta S\%$ of 1%, which is expressed by the following equation;

$$\eta = P / (S_{volume} \times \Delta S\%) \tag{20}$$

where P is the energy consumption of the corona discharge, S_{volume} is the volume of sludge. The P under DC corona discharge is measured through a resistor, while under AC corona, the discharge is calculated on the basis of Q-V Lissajous diagram. The result is depicted in Figure 20. It can be seen that the η is in the order of 10^{-1} mW/ (mL·1%), and the AC corona discharge has the highest η of 0.27 mW/ (mL·1%). Such a value is more than twice the value of η under negative DC corona discharge, and is almost 25 times higher than that under positive DC corona discharge. Clearly the positive DC corona has the lowest η.

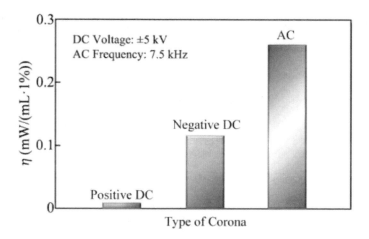

Figure 20. The energy consumption ratio induced by various types of corona discharge.

3.2.2. Effects of Treating Temperature and Treating Time on the WAS Pretreatment Performance with DC Corona Discharge

With the purpose of further understanding the pretreatment performance of DC corona discharge, experiments have been carried out to estimate the treating temperature and the treating time on the release of organic matter into the WAS. The sample used in this Section is SS II. As stated in Section 3.2.1, with the application of DC voltage between the multi-needle to plane electrode, corona discharge can be initiated with the generation of many oxidative species, energetic particles and UV irradiation, by which the disruption of microbial cells within the sludge can be achieved. The possible mechanism of the DC corona discharge has been discussed by measuring the discharge current through a 1 kΩ-resistor assembled with the grounding wire. The DC voltage applied here is ±4 kV [52]. The discharge current under positive DC corona is depicted in Figure 21(a). The current pulse of several mA with the frequency of tens of kHz are superimposed on the steady-state current (~10 μA) which can be recognized as the feature of streamer corona that generates cold plasma [53]. On the other hand, the discharge current waveform of negative corona is quite different, which is composed of a steady-state DC current and tens of microamperes impulses with the frequency of hundreds of kHz, as shown in Figure 21(b). This behavior should be attributed to the Trichel impulses [54]. Both the Trichel impulses and the positive streamer corona are associated with the ionization and the movement of space charge in air, which can introduce oxidative species thus disrupting the microbial cell within the WAS sample.

The relationship between the $\Delta S\%$ and the treating time under ±4 kV at 20^0C is shown in Figure 22. The $\Delta S\%$ exhibits obvious polarity dependence. With the positive corona, the $\Delta S\%$ is 13.3% at the treating time of 5 min, which gradually decreases to the value below zero with the treating time. For the negative corona, the $\Delta S\%$ first increases from 7.5% to 25.8% as the time varies from 5 min to 10 min. But a further increase in the time results in the reduction of the $\Delta S\%$ to -57.3% at 60 min. Such a test result indicates that the DC corona pretreatment possesses good performance with a short period of time.

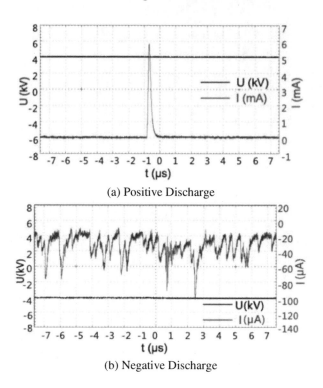

(a) Positive Discharge

(b) Negative Discharge

Figure 21. DC corona discharge current waveforms under ±4 kV in atmospheric air [52].

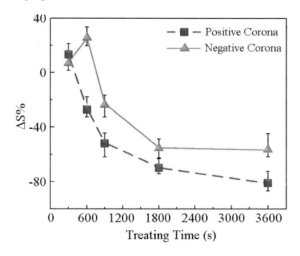

Figure 22. Effect of the treating time on the *ΔS%* under positive and negative corona at 20℃.

The reason for above result can be attributed to the "over oxidation effect" mentioned in the previous section. For the positive corona, the $\Delta S\%$ tends to be lower than zero as the time is longer than 10 min. This is mainly because fewer microbial cells in the WAS are survived with the corona exposure and more released organic matter is decomposed with the time. As clearly demonstrated in Section 3.2.1, the active particles and the oxidative species are more accessible for the WAS sample with positive corona, which results in further oxidation of organic matter and, therefore, the decrease of SCOD. With the lapse of the treating time, more oxidative species are generated that can reach the sludge surface to decompose the organic matter, therefore the $\Delta S\%$ decreases to a value even below zero. But for the negative corona, less active particles and oxidative species are injected into the sludge sample such that the over oxidation process doesn't play any remarkable role in reducing the SCOD when the treating time is shorter than 10 min. When the treating time is longer than 15 min, the over oxidation effect gradually becomes dominant and the $\Delta S\%$ tends to become a negative value. It is then proposed that the over oxidation effect would occur when the treating time reaches a certain value under both positive and negative corona, short time pretreatment with DC corona is beneficial for the release of organic matter and the following biodegradation process.

However, too short a time may not be available for the pretreatment of the WAS. It has been demonstrated that the OH radicals are the most reactive particles among oxidative particles generated by DC corona discharge, and it is also the main reactant of the oxidation process. The chemical reactions of OH radicals are free radical reactions in nature where the reaction rate is high. The rate constants for C-H and C-C bonds are mostly at 10^9 L/ (mol·s), which are close to the diffusion rate limit of 10^{10} L/ (mol·s). The chemical reaction time is less than 1 s and the biochemical reaction time was 1 s~10 s [55]. Nevertheless, the oxidative species generated in the corona need time to move deep into the sludge, otherwise the oxidation reaction may take place mainly on the sludge surface layer. Accordingly, the treating time in the range of several minutes is needed to show up the effect of those oxidative species introduced by the corona discharge.

Besides the change in SCOD, the variations in the NH_3-N, the total phosphorus and the pH value in response to the treating time have been estimated as well. Figure 23 exhibits a typical relationship between the $\Delta N\%$ and the treating time under ±4 kV at 20 ℃. The $\Delta N\%$ shows similar tendency with the time as compared with the $\Delta S\%$. For the positive corona, the $\Delta N\%$ monotonously decreases and is lower than zero as the time is longer than 15 min. For the negative corona, the $\Delta N\%$ increases as the treating time extends from 5 min to 10 min, but decreases as the time goes further to 60 min. The mechanisms are as follows. The proteins as the basic components of microbial cells contain a number of elements such as carbon, hydrogen, oxygen as well as nitrogen. The change in NH_3-N in the sludge is therefore related to the variation of protein content [14]. With the occurrence of over oxidation, the decomposition of proteins in the sludge occurs, leading to the reduction in the NH_3-N through the following ways [56]:

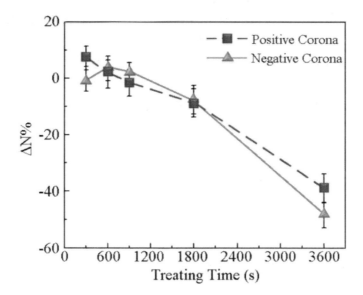

Figure 23. Typical relationship between the $\Delta N\%$ and the treating time under ± 4 kV at 20℃.

$$2NH_3 + 4O_2 \rightarrow NH_4NO_2 + H_2O_2 + 4O_2 \tag{21}$$

$$NH_4NO_2 + H_2O_2 \rightarrow NH_4NO_3 + H_2O \tag{22}$$

In addition, the ozone, O atoms and OH radicals can oxidize NH_3 to form N_2, N_2O and HNO_3. Then HNO_3 may form NH_4NO_3 in sludge [57]. With the formation of NO_3^-, N_2 and N_2O, the content of NH_3-N in the WAS sample becomes decreased. Another possible reason for the decrease in the $\Delta N\%$ can be considered from the viewpoint of the biochemical process. With the release of intracellular organic matter, carbon and nitrogen are provided to the microorganisms in the WAS for reproducing [58]. Such a process will consume nitrogen element, resulting in the reduction of the NH_3-N in the WAS. Moreover, the nitrification in sludge possibly oxidizes the ammonia into NO_3^- [59], by which the decrease in NH_3-N can be obtained.

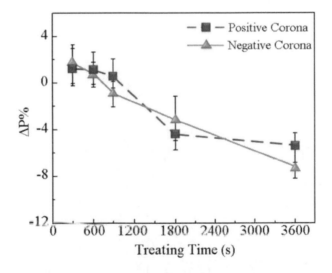

Figure 24. Typical change in the $\Delta P\%$ caused by the treating time at 20^0C.

With respect to the $\Delta P\%$, Figure 24 depicts a typical variation in the $\Delta P\%$ as a function of the treating time. For the positive and the negative corona, the $\Delta P\%$ decreases when the treating time elapses from 5 min to 60 min. The phosphorus element can exist with both its organic and inorganic forms in the wastewater. However, the proportion of organic phosphorus in wastewater is only ~20% of the total phosphorus. Additionally, the phosphorus just occupies a small fraction of the total weight of organic matter, which results in the smaller reduction of $\Delta P\%$ from 2% to -5%~-7% under positive and negative corona. Such a decrease in $\Delta P\%$ should be ascribed to the assimilation of organisms, by which the phosphorus as a nutrient donor is consumed [60].

Figure 25 gives a typical variation in pH value of the WAS sample with the treating time. Since the initial pH values for the un-treated WAS sample are slightly different, the relative pH value (ratio to initial value) has been employed here to estimate the treating time effect. It can be seen that the relative pH values in both positive and negative corona treated samples increase slightly. It should be attributed to the introduction of hydroxyl ions from the corona. Basically, the increase in the pH value is small, hence it will not seriously influence the following biodegradation process in WAS.

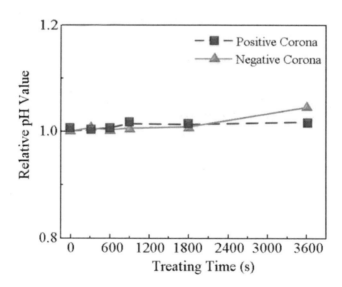

Figure 25. The change in relative pH as a function of the treating time under ±4 kV at 20°C.

From the results shown in Figures 22 to 25, it can be concluded that $\Delta S\%$, $\Delta N\%$ and $\Delta P\%$ decreases to a value even below zero when the DC corona treating time is over a certain value which suggests that the content of released organic matter for the following biodegradation process is reduced, hence the pretreatment efficiency gets worse. It is therefore proposed that shorter treating time with DC corona will lead to better pretreatment efficiency.

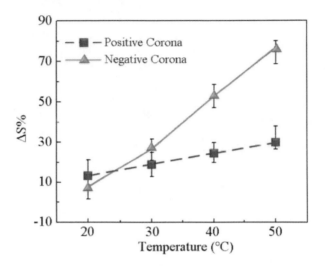

Figure 26. The temperature dependence of the $\Delta S\%$ with the treating time of 5 min.

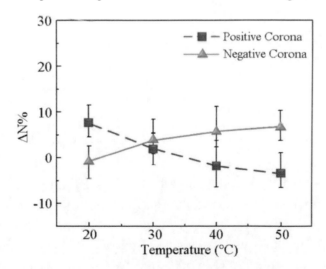

Figure 27. Influence of the temperature on the $\Delta N\%$ with the treating time of 5 min.

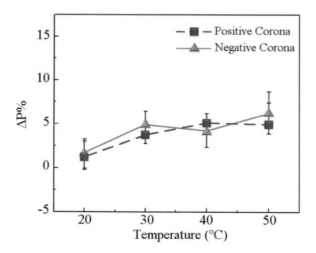

Figure 28. Influence of the temperature on the $\Delta P\%$ with the treating time of 5 min.

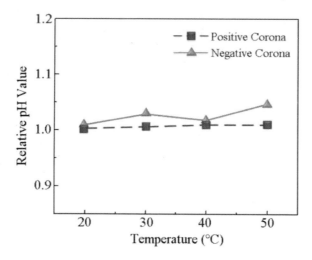

Figure 29. The temperature dependence of the relative pH value with the treating time of 5 min.

The other important factor that will be discussed in this section is the treating temperature. The temperature has a significant influence on the pretreatment performance as well. Figure 26 shows the temperature dependence of the $\Delta S\%$ with the treating time of 5 min. With the increase of the temperature from 20°C to 50°C, the $\Delta S\%$ obviously increases. The $\Delta S\%$ grows from 13.3% to 29.5% with positive corona, and from 7.5% to

76.7% with negative corona. The enhancement in the $\Delta S\%$ is more significant with negative corona than with positive corona. As stated earlier in Section 3.1, the phase transitions could occur on phospholipid foam bilayers of cell membranes with the growth of temperature [38].

Accordingly, the movement of cell membrane molecules around the phase transitions region becomes more active under higher temperature, hence the disruption of cell membranes is easier to be induced. Therefore, the pretreatment performance is improved by the high temperature. For the positive corona, more reactive species are capable of reaching the sludge as compared with the negative corona where the over oxidation is considered to limit the increase of $\Delta S\%$. As a result, the $\Delta S\%$ increases slowly under positive corona and becomes lower than under negative corona.

Figure 27 exhibits the influence of the temperature on the $\Delta N\%$ with the treating time of 5 min. When the temperature changes from 20^0C to 50^0C, the $\Delta N\%$ with negative corona increases from -0.7% to 6.8% whereas it reduces slightly from 7.5% to -3.4% with positive corona. Cell microstructure disruption tends to occur easily with the temperature, the release of intracellular organic matter results in the enhancement of $\Delta N\%$ with negative corona. For the positive corona, the over oxidation in the WAS leads to the decrease in $\Delta N\%$. Generally speaking, the change in $\Delta N\%$ is not so remarkable with the temperature, and negative corona pretreatment would have a positive effect due to the deficiency of over oxidation.

Different from the behaviors observed for the temperature dependent $\Delta N\%$, the $\Delta P\%$ slightly increases with the temperature with both positive and negative corona, approximately from 2% to 5%, as shown in Figure 28. Since the phosphorus from the microbial cells contributes to just a small part of the total phosphorus in the sludge sample, the disruption of cells has little effect on the enhancement of the $\Delta P\%$. The change in the relative pH value with the temperature is illustrated in Figure 29. The relative pH value doesn't show a remarkable growth in response to the temperature. Negative corona has a comparatively stronger effect on the pH value as compared to the positive corona. It is mainly because that the $\Delta N\%$ increases with temperature, hence the sludge sample tends to be alkaline in nature. For the positive corona pretreatment, the variation in the pH value is almost negligible which should be ascribed to the over oxidation induced decrease in the $\Delta N\%$. From the results shown in Figures 26 to 29, it is concluded that the temperature has a positive effect on the pretreatment performance, therefore better pretreatment efficiency can be achieved with higher temperature. Furthermore, the increase in the $\Delta S\%$ with the temperature is more remarkable than the $\Delta N\%$ and the $\Delta P\%$. This is helpful to form a proper content ratio among C: N: P in the WAS, hence the following biodegradation process could be encouraged.

3.3. WAS Pretreatment with PEF-Corona Discharge Combined Method

We have examined the role of the PEF and the corona discharge on the pretreatment of WAS. It is clear that by selecting proper working parameters, good pretreatment performance could be achieved. Further, if the PEF is associated with the corona discharge, should the combined method present some potential in the pretreatment with better results? In this section, the PEF-corona discharge combined method will be discussed and the pretreatment performance will be estimated.

In the combined method, the treating procedure used in the experiment is as follows. The WAS sample (SS I) is initially exposed to the DC corona discharge with the magnitude from ±2 kV to ±5 kV for 5 min. Afterwards, the sample is PEF treated with a magnitude of 10 kV/cm to 30 kV/mm for 10 min at 40^0C. In order to show the advantage of PEF-corona discharge combined method, the typical PEF pretreatment result is shown first as a comparison in Figure 30, where the $\Delta S\%$ grows from 8.8% to 28.7% with the PEF magnitude from 10 kV/cm to 30 kV/cm. Such behavior is well in agreement with that presented in Section 3.1. The WAS pretreatment performance under PEF-corona discharge combined method is shown in Figure 31. The result with a PEF of 0 kV/cm (DC corona works only) is also presented in the figure for comparison. When PEF is combined with positive corona, the $\Delta S\%$ exhibits a rise-and-fall tendency and the best pretreatment result occurs at 4 kV. For PEF combined with negative corona, the $\Delta S\%$ increases with the DC voltage. Such behavior should be attributed to the over oxidation phenomenon occurring with the positive voltage. As can be seen in Figures 30 and 31, the combined method obviously improves the pretreatment efficiency of PEF or DC corona. In addition, PEF combined with positive corona exhibits better performance than PEF combined with negative corona.

As the WAS sample is exposed to the DC corona, oxidative species like OH radicals, ozone et al. will be injected into the sample. This leads to the increase in SCOD even before the PEF application. There is another effect of the corona which assists the work of the PEF followed. The electric charges generated during the corona would become conductive particles as they are transported into the sludge, which increases the sludge conductivity. A conductivity measurement is carried out before and after the treatment of DC corona by a conductivity meter (Model-SC82, YOKOGAWA), the results are shown below in Figure 32. The conductivity becomes enhanced after corona charging, and increases with the DC voltage. The enhancement of the conductivity has an influence on the electroporation behavior.

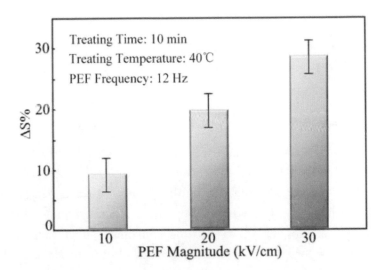

Figure 30. A typical effect of the PEF magnitude on the $\Delta S\%$.

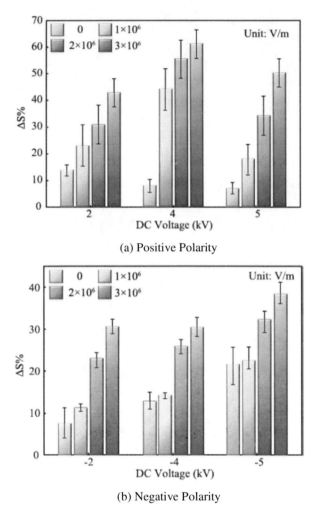

(a) Positive Polarity

(b) Negative Polarity

Figure 31. Influence of the DC voltage on the $\Delta S\%$ at various magnitudes of PEF.

It has been revealed that the time constant for membrane charging τ_m can be expressed by [16], where σ_i, σ_m and σ_e, are the conductivities of the cytoplasm, cell membrane and extracellular solution, respectively. ε_m is the dielectric permittivity of the membrane, d is the membrane thickness. When the σ_e increases, the charging time constant τ_m decreases. The cell membrane electroporation under PEF is more likely to occur. Accordingly, the increase in conductivity assists the microbial cell disruption under PEF thus a better performance will be achieved with the PEF-corona combined method.

$$\tau_m = \frac{R_{cell}\varepsilon_m}{2d\dfrac{\sigma_i\sigma_e}{\sigma_i+2\sigma_e} + R_{cell}\sigma_m}$$

(21)

One thing that should be brought to mind is that although the conductivity of the sludge gets increased by the corona discharge, it can't keep that high value constantly. Instead, the conductivity will decay with time as the charges within the sludge migrate toward the grounded electrode to reduce the charge number. Figure 33 indicates the decay behavior of the conductivity treated with both positive and negative corona. For the sludge treated with negative corona, the conductivity reduces from 6.6 mS/cm to 4.2 mS/cm in 12 min. For the positive corona-treated sludge, such value decreases from 6.3 mS/cm to 3.2 mS/cm. However, the conductivity of the corona-charged sludge at 12 min is still larger than that of the untreated sludge with 2.6 mS/cm. This suggests that the lifetime of the charges implanted is long enough to assist the 10 min-PEF treatment performed, but the conductivity may reduce to a smaller value, hence its effect on encouraging the membrane electroporation is restricted.

The polarity effect of the combined treatment method can be understood from the viewpoint of charge transportation within the WAS sample. The charges injected into the sludge are distributed non-uniformly and are mainly concentrated close to the surface. On the other hand, the concentration at the bottom is relatively lower. As shown in Figure 34, when positive charges are deposited, they can migrate to the grounded electrode under the PEF thanks to the driven field generated with their identical polarities. The positive charges are therefore dispersing widely in the sludge, as depicted in Figure 34(a). As a result, the positive charges in a relatively larger volume of the WAS will encourage the PEF induced cell membrane electroporation. For the negative charges, as their polarity is the opposite of the pulsed voltage, they are attracted to accumulate in a shallow layer of the sludge surface, as shown in Figure 34(b). The effect of negative charges on the PEF induced membrane electroporation is thereby limited. In short, the positive charge would contribute much to the PEF treatment.

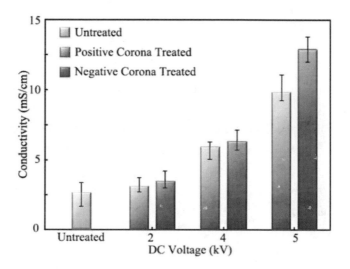

Figure 32. Conductivity of the WAS sample as a function of the DC voltage.

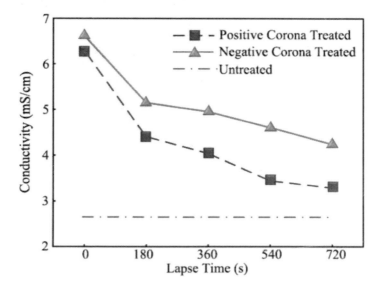

Figure 33. Decay of conductivity of the WAS sample.

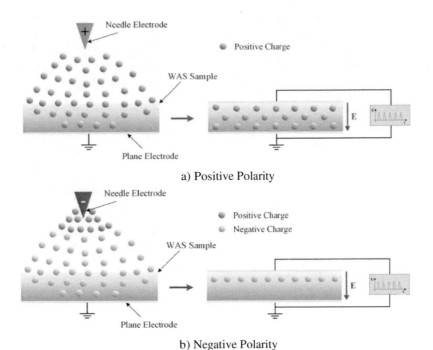

a) Positive Polarity

b) Negative Polarity

Figure 34. Charge distribution in the WAS with corona discharge of different polarities.

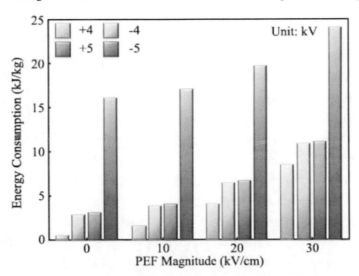

Figure 35. Energy consumption caused by various treatment conditions with PEF and DC corona discharge combined method.

Although the combined method shows an obvious improvement of the pretreatment efficiency as compared with either the PEF or the DC corona discharge, the energy consumption of such a method is evidently higher. Figure 35 shows the energy consumption of PEF-corona discharge combined method with various PEF magnitudes and DC voltages. The energy consumed by the pretreatment is basically in the range of several to tens of kJ/kg. Clearly, with the increase of PEF magnitude or DC voltage, the energy consumption appears to increase. In addition, the combined method presents higher energy consumption than that of either one. The positive DC corona will consume more energy as compared to the negative DC corona. In short, from the viewpoint of the industrial application of PEF-corona discharge combined method, the energy consumption must be taken into deep account and a balance should be made between the performance efficiency and the energy consumption.

CONCLUSION

In this chapter, the WAS pretreatment methods based on PEF, corona discharge and their combination have been introduced. The factors that influence the pretreatment performance have been estimated and the possible mechanisms have been discussed. Generally speaking, with proper selection of treating parameters, all three methods have good potential in industrial applications. Further, a balance should be made between the pretreatment efficiency and the energy consumption, by which the method would be beneficial from both technical and economical points of view.

The main conclusions can be summarized as follows:

1. PEF as an effective method for the WAS pretreatment is dependent upon a number of parameters. With the increase of the PEF magnitude and temperature, the *IOMRI* that represents the pretreatment efficiency increases exponentially, which indicates that the PEF pretreatment holds better performance at higher fields and temperature. However, the treating time has much less impact on the *IOMRI*. The *IOMRI* decreases as the PEF

frequency grows from 1 Hz to 100 Hz. A suitable frequency should be selected for an expected pretreatment efficiency.

2. Corona discharge is another method showing strong dependence of the pretreatment performance upon the treating parameters like voltage type, treating temperature and time, frequency and polarity. For the DC corona discharge, the pretreatment performance exhibits a polarity dependence. With respect to positive corona, the pretreatment efficiency decreases with the voltage. However, for the negative corona, the efficiency increases. For the AC corona discharge, the efficiency is dependent upon the AC frequency. At a frequency of 5 kHz, the $\Delta S\%$ decreases monotonously with the increase of AC magnitude. At a frequency of 7.5 kHz, the $\Delta S\%$ initially decreases but then tends to increase with the magnitude. At a frequency of 10 kHz, the $\Delta S\%$ obviously increases with the magnitude. The difference in the pretreatment efficiency caused by the polarity or AC frequency is associated with the over oxidation effect generated mainly by the positive corona discharge. In addition, the energy consumption ratio is in the order of 10^{-1} mW /(mL·1%), the consumption caused by the AC voltage is higher than the DC voltage.

The pretreatment performance also depends upon the temperature and the time. With the lapse of treating time from 5 min to 60 min, the $\Delta S\%$ under both positive and negative corona basically decreases. The reduction with positive corona is larger than that with negative corona. Meanwhile, the $\Delta N\%$ and the $\Delta P\%$ decrease. However, the pH value slightly increases. With the temperature grows from 20℃ to 50℃, the $\Delta S\%$ increases both under positive and negative corona. The increment under negative corona is much larger than that under positive corona. The $\Delta N\%$ of WAS increases under negative corona but decreases under positive corona. The $\Delta P\%$ shows an increasing trend under both positive and negative corona although its increment is quite small. Moreover, the pH value exhibits a negligible variation. Such variations in the $\Delta S\%$, the $\Delta N\%$, the $\Delta P\%$ and the pH value with the temperature facilitate the formation of a proper content ratio among C: N: P that is helpful for the following biodegradation process.

3. The PEF-corona discharge combined pretreatment method clearly shows better performance than either of them, and is dependent upon the polarity and the magnitude of the DC voltage. For the PEF-positive corona combination, the pretreatment performance initially increases as the DC voltage grows from 2 kV to 4 kV but decreases as the voltage rises to 5 kV. For the PEF-negative corona combination, the performance increases with the voltage. PEF-positive corona combined pretreatment exhibits better efficiency than the PEF-negative corona. The energy consumption of the PEF-corona discharge combined pretreatment is higher than that with either of them, but is still in the comparable range of several to tens of kJ/kg. The oxidative species and electric charges introduced by the corona are considered to be the very reason encouraging the PEF pretreatment. The oxidative species help to disrupt the cell microstructure, while the electric charges can improve the sludge conductivity, hence the membrane electroporation is facilitated.

ACKNOWLEDGMENTS

The authors would like to thank the financial support from Research Institute Of Tianjin Capital Environment Protection Group Company Limited, Tianjin, China, who launched a project aimed at "Developing a WAS pretreatment method based on high voltage technique" and gave us great suggestions on experimental design and data analysis.

Dr. Yu Gao would like to sincerely appreciate the hardworking of his postgraduate students, Mr. Yekun Men, Mr. Yongdi Deng and Mr. Ning Zhao, who have been engaged in the project and spent almost all of their personal time on developing the WAS pretreatment method based on the PEF and the corona discharge plasma. Dr. Yu Gao is also grateful for the contribution from Mr. Ziyi Li who contributed much to the design of photographs used in this chapter.

This work is partly supported by the National Nature Science Foundation of China (NSFC 51677127, 51677128, 51707132) and the Natural Science Foundation of Tianjin (16JCQNJC06800).

REFERENCES

[1]   Wang, M. Y., Webber, M., Finlayson, B. L. and Barnnet, J. 2008. "Rural industries and water pollution in China." *Journal of Environmental Management* 86:648-59.

[2]   Xiao, Y. T. and He. D. W. 2015. *Technology of municipal wastewater treatment.* Beijing: China Building Materials Press.

[3]   Eskicioglu, C., Prorot, A., Marin, J., Droste, R. L. and Kennedy, K. J. 2008. "Synergetic pretreatment of sewage sludge by microwave irradiation in presence of H2O2 for enhanced anaerobic digestion." *Water Research* 42: 4674-82.

[4]   Xiao, B. Y., Liu, C., Liu, J. X. and Guo, X. S. 2015. "Evaluation of the microbial cell structure damages in alkaline pretreatment of waste activated sludge." *Bioresource Technology* 196:109-15.

[5]   Appels, L., Baeyens, J., Degrève, J. and Dewil, R. 2008. "Principles and potential of the anaerobic digestion of waste-activated sludge." *Progress in Energy and Combustion Science* 34:755-81.

[6]   Pilli, S., Bhunia, P., Song, Y., LeBlanc, R. J., Tyagi, R. D. and Surampalli, R. Y. 2011. "Ultrasonic pretreatment of sludge: A review." *Ultrasonics Sonochemistry* 18:1-18.

[7]   Mottet, A., François, E., Latrille, E., Steyer, J. P., Deleris, S., Vedrenne, F. and Carrere, H. 2010. "Estimating anaerobic biodegradability indicators for waste activated sludge." *Chemical Engineering Journal* 160:488-96.

[8]   Burger G, and Parker, W. 2013. "Investigation of the impacts of thermal pretreatment on waste activated sludge and development of a pretreatment model." *Water Research* 47:5245-56.

[9]   Pei, J., Yao, H., Wang, H., Shan, D., Jiang, Y. C., Ma, L. Q. Y. and Yu, X. H. 2015. "Effect of ultrasonic and ozone pre-treatments on pharmaceutical waste activated sludge's solubilization, reduction, anaerobic biodegradability and acute biological toxicity." *Bioresource Technology* 192:418-23.

[10]   Arena, C. B., Sano, M. B., Nichole, R. M. and Davalos, R. V. 2011. "Theoretical considerations of tissue electroporation with high-

frequency bipolar pulses." *IEEE Transactions on Biomedical Engineering* 58:1474-82.

[11] Jin, T. Z., Guo, M. M. and Zhang, H. Q. 2015. "Upscaling from benchtop processing to industrial scale production: More factors to be considered for pulsed electric field food processing." *Journal of Food Engineering* 146:72-80.

[12] Sawayama, S., Inoue, S., Tsukahara, K. and Ogi, T. 1996. "Thermochemical liquidization of anaerobically digested and dewatered sludge and anaerobic retreatment." *Bioresource Technology* 55:141-4.

[13] Penaud, V. Delgenès, J. P. and Moletta, R. 1999. "Thermo-chemical pretreatment of a microbial biomass: influence of sodium hydroxide addition on solubilization and anaerobic biodegradability" *Enzyme and Microbial Technology* 25:258-63.

[14] Nah, I. W., Kang, Y. W., Hwang, K. Y. and Song, W. K. 2000, "Mechanical pretreatment of waste activated sludge for anaerobic digestion process." *Water Research* 34:2362-8.

[15] Choi, H., Jeong, S. W. and Chung, Y. J. 2006. "Enhanced anaerobic gas production of waste activated sludge pretreated by pulse power technique." *Bioresource Technol* 97:198-203.

[16] Kotnik, T., Kramar, P., Pucihar, G., Miklavcic, M. and Tarek, M. 2012. "Cell membrane electroporation-Part 1: The phenomenon." *IEEE Electrical Insulation Magazine* 28:14-23.

[17] Islamov, R. S. and Krishtafovich, Y. A. 2011. "A peculiarity of silver-based corona wire heating on ozone generation." *Journal of Electrostatics* 69:236-9.

[18] Koners, U., Heinz, V., Knorr, D., Loffler, M. and Schmidt, W. 2006. "Effects of pulsed electric field (PEF) application on activated wastewater treatment sludge." In *1st Euro-Asian Pulsed Power Conference (EAPPC06)*, Chengdu, September 18-22.

[19] Lee, J. S. and Chang, I. S. 2014. "Membrane fouling control and sludge solubilization using high voltage impulse (HVI) electric fields." *Process Biochemistry* 49:858-62.

[20] Salerno, M. B., Hyung-Sool, L., Prathap, P. and Rittmann, B. E. 2009. "Using a pulsed electric field as a pretreatment for improved biosolids digestion and methanogenesis." *Water Environment Research* 81:831-9.

[21] Lee, I. S. and Rittmann, B. E. 2011. "Effect of low solids retention time and focused pulsed pre-treatment on anaerobic digestion of waste activated sludge." *Bioresource Technology* 102:2542-8.

[22] Rittmann, B. E., Lee, H. S., Zhang, H., Alder, J., Banaszak, J. E. and Lopez, R. 2008. "Full-scale application of focused-pulsed pre-treatment for improving biosolids digestion and conversion to methane." *Water Science and Technology* 58:1895-901.

[23] Malik, M. A., Ghaffar, A. and Malik, S. A. 2001. "Water purification by electrical discharges." *Plasma Sources Science and Technology* 247:755–763.

[24] Korachi, M., Turan, Z., Şentürk, K., Sahin, F. and Aslan, N. 2009. "An investigation into the biocidal effect of high voltage AC/DC atmospheric corona discharges on bacteria, yeasts, fungi and algae." *Journal of Electrostatics* 67:678-85.

[25] Scholtz, V., Julak, J. and Stepankova, B. 2011. "The possibilities and comparison of point-to-plane and point-to-point corona discharge for the decontamination or sterilization of surfaces and liquids." *Plasma Medicine* 1:21-5.

[26] Machala, Z., Jedlovsky, I., Chladekova, L., Pongrac, B., Giertl, D. and Janda, M. 2009. "DC discharges in atmospheric air for bio-decontamination-spectroscopic methods for mechanism identification." *The European Physical Journal D* 54:195-204.

[27] Matveev, I. B., Serbin, S. I. and Washchilenko, N. V. 2014. "Sewage Sludge-to-Power." *IEEE Transactions on Plasma Science* 42:3876-80.

[28] Curds, C. R. 1988. "Protozoa in biological sewage treatment processes." *Water Research* 14:225-36.

[29] Wagner, M., Erhart, R., Manz, W., Amann, R., Lemmer, H., Wedi, D. and Schleifer, K. H.1994. "Development of an rRNA2 targeted oligonucleotide probe specific for the genus Acinetobacter and its

application for in situ monitoring in activated sludge." *Applied and Environmental Microbiology* 60:792-800.

[30] Henze, M., Ekama, G. A. and Brdjanovic, D. 2011. *Biological wastewater treatment: principles, modelling and design.* London: IWA.

[31] Weaver, James C., Smith, Kyle C., Esser, Axel T., Son, Reuben S. and Gowrishankar, T. R. 2012. "A brief overview of electroporation pulse strength-duration space: a region where additional intracellular effects are expected." *Bioelectrochemistry* 87:236-43.

[32] Weaver, James C. 2003. "Electroporation of biological membranes from multicellular to nan scales." *IEEE Transactions on Dielectrics and Electrical Insulation* 10:754-68.

[33] Y. L. Gao, and D. Ma. 2006. *New technology for wastewater biological treatment.* Beijing: China Building Materials Press.

[34] Bilska, A. O., De Bruin, K. A. and Krassowska, W. 2000. "Theoretical modeling of the effects of shock duration, frequency, and strength on the degree of electroporation." *Bioelectrochemistry and Bioenergetics* 51:133-43.

[35] Frey, W., White, J. A., Price, R. O., Blackmore, P. F., Joshi, R. P., Nuccitelli, R. Beebe, S. J., Schoenbach, K. H. and Kolb, J. F. 2006. "Plasma membrane voltage changes during nanosecond pulsed electric field exposure." *Biophysical Journal* 90:3608-15.

[36] Talele, S., Gaynor, P., Cree, M. J. and Ekeran, J. 2010. "Modelling single cell electroporation with bipolar pulse parameters and dynamic pore radii." *Journal of Electrostatics* 68:261-74.

[37] Hu, Q., Joshi, R. P. and Schoenbach, K. H. 2005. "Simulations of nanopore formation and phosphatidylserine externalization in lipid membranes subjected to a high-intensity, ultra short electric pulse." *Physical Review E: Statistical, Nonlinear & Soft Matter Physics* 72:031902.

[38] Exerova, D. and Nikolova, A. 2002. "Phase transaction in phospholipid foam bilayers." *Langmuir* 8:3102-8.

[39] Heimburg, T. 2000. "Monte Carlo simulations of lipid bilayers and lipid protein interactions in the light of recent experiments." *Current Opinion in Colloid and Interface Science* 5:224-31.

[40] Lebovka, N. I., Praporscic, L., Ghnimi, S. and Vorobiev, E. 2005. "Temperature enhanced electroporation under the pulsed electric field treatment of food tissue." *Journal of Food Engineering* 69(2):177-84.

[41] Zhu, Y. M., Zhang, M. X. and Sun, P. H. 2005. "Effects of needle radius on current voltage characteristics of multi-needle-to-plate corona discharge." *Transactions of Beijing Institute of Technology* 25:37-140.

[42] Yamatake, A., Fletcher, J., Yasuoka, K. and Ishii, S. 2006. "Water treatment by fast oxygen radical flow with DC-driven microhollow cathode discharge." *IEEE Transactions on Plasma Science* 34:1375-81.

[43] Zhang, Y., Zhou, M. H. and Lei, L. C. 2007. "Degradation of 4-chlorophenol in different gas–liquid electrical discharge reactors." *Chemical Engineering Journal* 132:325-33.

[44] Das-Gupta, D. K. 1990. "Decay of electrical charges on organic synthetic polymer surfaces." *IEEE Dielectrics and Electrical Insulation* 25:503-8.

[45] Hagve, T. A. 1988. "Effects of unsaturated fatty acids on cell membrane functions." *Scandinavian Journal of Clinical & Laboratory Investigation* 48:381-8.

[46] Smith, F. A. and Smith, S. E. 1989. "Membrane transport at the biotrophic interface. An overview." *Functional Plant Biology* 16:33-43.

[47] Laroussi, M. 2005. "Low temperature plasma-based sterilization: overview and state-of-the-art." *Plasma Process Polymers* 2:391-400.

[48] Joshi, A. A., Locke, B. R., Arce, P. and Finney, W. C. 1995. "Formation of hydroxyl radicals, hydrogen peroxide and aqueous electrons by pulsed streamer corona discharge in aqueous solution." *Journal of Hazardous Materials* 41:3-30.

[49] Gao, Y., Deng, Y. D. and Men, Y. K. 2016. "Disruption of microbial cell within waste activated sludge by DC corona assisted pulsed electric field." *IEEE Transactions on Plasma Science* 44:2682-91.

[50] Yan, K., Yan, H. X., Cui, M., Miao, J. S., Wu, X. L., Bao, C. G. and Li, R. N. 1998. "Corona induced non-thermal plasmas: Fundamental study and industrial applications." *Journal of Electrostatics* 44:17-39.

[51] Gao, Y., Zhao, N., Men, Y. K. and Deng, Y. D. 2017. "Disruption of urban excess sludge based on corona discharge." *High Voltage Engineering* 43:2666-72.

[52] Gao, Y., Zhao, N., Deng, Y. D., Wang, M. H. and Du, B. X. 2018. "Effects of treatment time and temperature on the DC corona pretreatment performance of waste activated sludge." *Plasma Science and Technology* 20:1-10.

[53] Timoshkin, I. V., Maclean, M., Wilson, M. P., Given, M. J., MacGregor, S. J., Wang, T. and Anderson, J. G. 2012. "Bactericidal effect of corona discharges in atmospheric air." *IEEE Transactions on Plasma Science* 40:2322-33.

[54] Fridman, A., Chirokov, A. and Gutsol, A. 2005. "Non-thermal atmospheric pressure discharges." *Journal of Physics D: Applied Physics* 38:1-24.

[55] Yu, S. J. 2010 "The specifity of hydroxyl Free radical and the methods to detect it." *Guangdong Chemical Industry*, 37:141-3.

[56] Tanthapanichakoon, W., Charinpanitkul, T., Chaiyo, S., Dhattavorn, N., Chaichanawong, J., Sano, N. and Tamon, H. 2004. "Effect of oxygen and water vapor on the removal of styrene and ammonia from nitrogen by non-pulse corona-discharge at elevated temperatures." *Renewable and Sustainable Energy Reviews* 97:213-23.

[57] Shvedchicov, A. P., Belousova, E. V., Polyakova, A. V., Ponizovsky, A. Z. and Ponizovsky, L. Z. 1996. "Oxidation of ammonia in moist air by use of pulse corona discharge technique." *Radiation Physics & Chemistry* 47:475-7.

[58] Vlaeminck, S. E., Hay, A. G., Maignien, L. and Verstraete, W. 2011. "In quest of the nitrogen oxidizing prokaryotes of the early Earth." *Environmental Microbiology* 13:283-95.

[59] Morales, G., Sanhueza, P. and Vidal, G. 2015. "Effect of the carbon source on nitrifying in an activated sludge system treating aquaculture wastewater." *Journal of Agricultural Science and Technology* 7:36-44.

[60] Hauduc, H., Rieger, L., Oehmen, A., Loosdrecht, M. C., Comeau, Y., Héduit, A., Vanrolleghem, P. A. and Gillot, S. 2013. "Critical review of activated sludge modeling: State of process knowledge, modeling concepts, and limitations." *Biotechnology & Bioengineering* 110:24-46.

BIOGRAPHICAL SKETCHES

Tao Han

Affiliation: School of Electrical and Information Engineering, Tianjin University

Research and Professional Experience: He received the ME and PhD degrees in electrical engineering from Tianjin University, China, in 2012 and 2015, respectively. Since 2015, he has been a lecturer at School of Electrical and Information Engineering in Tianjin University

Professional Appointments: His main research interests are degradation of cable insulation and partial discharge detection.

Publications from the Last 3 Years:
1. B. X. Du, J. G. Su, T. Han, Compressive stress dependence of electrical tree growth characteristics in EPDM, *IEEE Transactions on Dielectrics and Electrical Insulation*, 2018, 25 (1), 13-20.

2. B. X. Du, J. G. Su, T. Han, Effects of Mechanical Stretching on Electrical Treeing Characteristics in EPDM, *IEEE Transactions on Dielectrics and Electrical Insulation,* 2018, 25 (1), 84-93.

3. T. Han, B. X. Du, J. G. Su, Electrical Tree Initiation and Growth in Silicone Rubber under Combined DC-Pulse Voltage, *Energies,* 2018, 11 (4), 764.

4. B. X. Du, J. G. Su, T. Han, Temperature-Dependent Electrical Tree in Silicone Rubber under Repetitive Pulse Voltage, *IEEE Transactions on Dielectrics and Electrical Insulation,* 2017, 24 (4), 2291-2298.

5. B. X. Du, J. S. Xue, J. G. Su, T. Han, Effects of ambient temperature on electrical tree in epoxy resin under repetitive pulse voltage, *IEEE Transactions on Dielectrics and Electrical Insulation,* 2017, 24 (3), 1527-1536.

6. B. X. Du, J. G. Su, J. Li, T. Han, Effects of Mechanical Stress on Treeing Growth Characteristics in HTV Silicone Rubber, *IEEE Transactions on Dielectrics and Electrical Insulation,* 2017, 24 (3), 1547-1556.

7. B. X. Du, L. W. Zhu, T. Han, Effect of Low Temperature on Electrical Treeing of Polypropylene with Repetitive Pulse Voltage, *IEEE Transactions on Dielectrics and Electrical Insulation,* 2016, 23 (4), 1915-1923.

8. B. X. Du, M. M. Zhang, T. Han, L. W. Zhu, Effect of Pulse Frequency on Tree Characteristics in Epoxy Resin under Low Temperature, *IEEE Transactions on Dielectrics and Electrical Insulation,* 2016, 23 (1), 104-112.

9. Y. Yu, B. X. Du, J. X. Jin, T. Han, J. G. Su, Effect of Magnetic Field on Electrical Treeing Behavior of Silicone Rubber at Low Temperature, *IEEE Transactions on Applied Superconductivity,* 2016, 26 (7), 1-4.

10. T. Han, B. X. Du, Y. Yu, X. Q. Zhang, Effect of Cryogenic Temperature on Tree Characteristics in Silicone Rubber/ SiO_2Nanocomposites Under Repetitive Pulse Voltage, *IEEE Transactions on Applied Superconductivity,* 2016, 26 (7), 1-4.

Yong Liu

Affiliation: School of Electrical and Information Engineering, Tianjin University

Research and Professional Experience: He received ME and PhD degrees in electrical engineering from Tianjin University, China, in 2006 and 2009, respectively. From 2009 to 2015, he was a lecturer at School of Electrical and Information Engineering in Tianjin University. From 2014 to 2015, he worked as a research fellow with the NSERC/Hydro-Quebec/UQAC Industrial Chair on Atmospheric Icing of Power Network Equipment, Canada. From 2015 to now, he is an associate professor at School of Electrical and Information Engineering in Tianjin University.

Professional Appointments: His main research interests are ageing evaluation and performance monitoring of outdoor insulators.

Publications from the Last 3 Years:
1. J. Li, B. X. Du, J. G. Su, H. C. Liang, Y. Liu, Surface Layer Fluorination-Modulated Space Charge Behaviors in HVDC Cable Accessory, *Polymers,* 2018, 10(5), 500.
2. Y. Liu, Farzaneh M., B. X. Du, Investigation on Shed Icicle Characteristics and Induced Surface Discharges along a Suspension Insulator String during Ice Accretion, *IET Generation Transmission and Distribution*, 2017, 11(5), 1265-1269.
3. Y. Liu, B. X. Du, Farzaneh M., Self-Normalizing Multivariate Analysis of Polymer Insulator Leakage Current Under Severe Fog Conditions, *IEEE Transactions on Power Delivery*, 2017, 32(3), 1279-1286.
4. Y. Liu, Farzaneh M., B. X. Du, Using chaotic features of leakage current for monitoring dynamic behavior of surface discharges on an ice-covered insulator, *IEEE Transactions on Dielectrics and Electrical Insulation*, 2017, 24(4), 2607-2615.

5. Y. Liu, Farzaneh M., B. X. Du, Nonlinear characteristics of leakage current for flashover monitoring of ice-covered suspension insulators, *IEEE Transactions on Dielectrics and Electrical Insulation.* 2016, 23(3), 1242-1250.

6. Y. Liu, B. W. Xia, B. X. Du, Farzaneh M., Influence of fine metal particles on surface discharge characteristics of outdoor insulators, *Energies,* 2016, 9(2), en9020087.

7. Y. Liu, D. Zhang, H. Xu, S. M. Ale-emran, B. X. Du, Characteristic analysis of surface damage and bulk micro-cracks of SiR/SiO$_2$ nanocomposites caused by surface arc discharges, *IEEE Transactions on Dielectrics and Electrical Insulation*, 2016, 23(4), 2102-2109.

8. B. X. Du, H. Xu, Y. Liu, Effects of wind condition on hydrophobicity behavior of silicone rubber in corona discharge environment, *IEEE Transactions on Dielectrics and Electrical Insulation,* 2016, 23(1), 385-393.

9. B. X. Du, A. Li, Y. Liu, Effect of direct fluorination on partial discharge characteristics of polyimide film used as magnet wire insulation of generator, *IET Generation Transmission and Distribution,* 2016, 10(9), 2251-2258.

10. Y. Liu, Z. L. Li, B. X. Du, Kinetics of Charge Accumulation and Decay on Silicone Rubber/SiO$_2$ Nanocomposite Surface, *Journal of Electrical Engineering and Technology*, 2016, 11(5), 1332-1336.

Yu Gao

Affiliation: Tianjin University, China

Research and Professional Experience: Dr. Yu Gao received the PhD degree from School of Electrical and Information Engineering, Tianjin University, China in 2009, then he joined in High Voltage and Electrical Insulation Lab in Tianjin University as a lecturer. In 2012, he was promoted as an associate professor in the field of high voltage and electrical insulation technology. From December 2017, he has been working as an Academic Visitor in the University of Southampton, UK. His research interests include ageing phenomena of polymer insulating materials, pulsed power technology and its application in environment engineering, overvoltage mechanism and protection for power systems. He has published over 30 peer-reviewed journal papers, more than 60 conference papers, two book chapters with respect to polymer ageing under radioactive environment as author and co-author. He is a member of IEEE DEIS and a member of DEIS Technical Committee on Nanodielectrics.

Professional Appointments: High Voltage and Electrical Insulation

Honors: "Most Valuable Poster Presentation" in ISEIM2017, Japan.

Publications from the Last 3 Years:
1. Y. Gao, Y. D. Deng, and Y. K. Men, "Disruption of microbial cell within waste activated sludge by DC corona assisted pulsed electric field," *IEEE Trans. Plasma Sci.*, vol. 44, no. 11, pp. 2682-2691, Jul. 2016.
2. Y. Gao, N. Zhao, Y. D. Deng, M. H. Wang, and B. X. Du, "Effects of treatment time and temperature on the DC corona pretreatment performance of waste activated sludge," *Plasma Sci. Technol.*, vol. 20, no. 2, pp. 025501-025510. Dec. 2018.
3. Y. Gao, J. L. Wang, F. Liu, and B. X. Du, "Surface potential decay of negative corona charged epoxy/Al_2O_3 nanocomposites degraded by 7.5 MeV-electron beam," *IEEE Trans. Plasma Sci.*, vol. 46, no. 7, pp. 2721-2729, Apr. 2018.
4. Y. Gao, F. Liu, J. L. Wang, X. F. Wang, and B. X. Du, "Improvement on partial discharge resistance of epoxy/Al_2O_3 nanocomposites by irradiation with 7.5 MeV electron beam," *IEEE Access,* vol. 6, pp. 25121-25129, Jan. 2018.
5. Y. Gao, Y. Q. Yuan, L. Y. Chen, J. Li, S. H. Huang, and B. X. Du, "Direct fluorination induced variation in interface discharge

behavior between polypropylene and silicone rubber under AC voltage," *IEEE Access,* vol. 6, pp. 23907-23917, Apr. 2018.

6. Y. Gao, J. Li, Y. Q. Yuan, S. H. Huang, and B. X. Du, "Trap distribution and dielectric breakdown of isotactic polypropylene/propylene based elastomer with improved flexibility for DC cable insulation," *IEEE Access,* vol. 6, pp. 58645-58661, Oct. 2018.

7. Y. Gao, M. H. Wang, N. Zhao, Z. Y. Li, and B. X. Du, "Research progress in surface charge characteristics of solid insulating materials," *High Voltage Engineering*, vol. 44, no. 8, pp. 2628-2645, Aug. 2018.

8. Y. Gao, S. H. Huang, L. Y. Chen, H. Xu, and B. X. Du, "Effect of elastomer content on interface discharge behavior between polypropylene and silicone rubber under AC voltage," *High Voltage Engineering,* vol. 44, no. 9, pp. 2874-2880, Sep. 2018.

9. Y. Gao, N. Zhao, Y. K. Men, and Y. D. Deng, "Disruption of urban excess sludge Based on Corona Discharge," *High Voltage Engineering,* vol. 43, no. 8, pp. 2666-2672, Aug. 2017.

10. Y. Gao, Y. K. Men, J. Zhang, F. J. Liu, Y. D. Deng, and Z. Wang, "Release behavior of organic matters from microorganism cells within excess sludge under microsecond pulsed electric field," *High Voltage Engineering*, vol. 42, no. 8, pp. 2675-2682, Aug. 2016.

In: The Activated Sludge Process
Editor: Benjamin Lefèbvre

ISBN: 978-1-53615-202-9
© 2019 Nova Science Publishers, Inc.

Chapter 2

MOLECULAR METHODS TO STUDY THE MICROBIAL PHENOMENA OF BULKING, FOAMING, AND ZOOGLEAL BULKING IN ACTIVATED SLUDGE

María Isabel Neria-González[*]
División de Ingeniería Química y Bioquímica,
Tecnológico de Estudios Superiores de Ecatepec,
Edo. de México, México

ABSTRACT

In this chapter, a description of the fundamental techniques in molecular biology for the analysis of the microbiota of activated sludge is given. Because activated sludge is a heterogeneous system of organisms, organic and inorganic material, givig a specific protocol for each of the molecular techniques would be imprudent. This is because even the same composition of the activated sludge generates a challenge in terms of proper handling and treatment of the samples. In addition, further barriers

---

[*] Corresponding Author Email: ibineria@hotmail.com, mineriag@tese.edu.mx.

include the type of equipment available in the laboratory and economic resources.

Microorganisms are fundamental in wastewater's biological treatment; the cleaning of water relies on the settleability of activated sludges. The term of activated sludge is commonly used to define a heterogeneous assembly of microorganisms organized in aggregates named "flocs." Aggregation refers mainly to bacterial production of exopolymers that promote attachment of microorganism and inert matter. However, filamentous organisms are an essential part of the floc population in an activated sludge process, forming the backbone to which floc-forming bacteria adhere. But, a high proliferation or lacking filamentous bacteria in the sludge can affect the operation conditions of a wastewater treatment plant due to the formation of pinpoint flocs, filamentous bulking, and viscous or zoogleal bulking. The pinpoint flocs are small and mechanically fragile, presenting low settleable properties due to the lack of a filamentous bacterial backbone. Filamentous bulking is generated by the filamentous bacteria overgrowth in the sludge, leading to poor sludge settleability, i.e., poor thickening characteristics of the sludge. Viscous bulking is caused by an excessive amount of extracellular polysaccharides (EPS), inducing a negative effect on the biomass thickening and compaction due to the water-retentive nature of EPS. It produces an activated sludge with a density closer to that of the surrounding water, increasing the sludge volume index, and, in some severe cases, there is no solids separation. In addition, some filamentous bacteria could induce foaming, which affects the operational conditions of the plant. These microbial phenomena can be summarized in three factors that impact the efficiency of the wastewater treatment process: filamentous bacteria, production of EPS, and floc-forming bacteria. The past decades have witnessed a significant growth in the use of optical and electron microscopy to analyze the development of activated sludges and identify filamentous bacteria. These techniques have increased the knowledge on the structural characterization of activated sludge, including morphological, physical, and chemical parameters that are closely related to solid–liquid separation. Image analysis methods hava been developed also to quantify the abundance of protruding filamentous bacteria based on filament length relative to floc area, but the activated sludge is not a mono-culture, a diversity of microorganisms exists with different morphological and physiological characteristics. Thus, molecular tools represent a big ally for further identifying filamentous bacteria, i.e., they allow determining phylogenetic microbial diversity in the activated sludge, identifying the microbial species and their abundance. Phylogenetic information can be the basis to determine the physiological status of activated sludge, recognizing which are metabolically active microbial populations, likewise, it allows monitoring of cultivation-independent phylogenetically-defined populations in the activated sludge, leading to a dynamic study of the wastewater depuration process.

Keywords: identification, quantification, bulking, molecular tools

INTRODUCTION

Currently, it is known that microbial communities are of great importance in the efficiency and robustness of a wastewater treatment plant. Thus, the microbial composition of these communities is related with the biological system performance to clean the wastewaters; the functions of each microbial population are essential to improve the control of the treatment system. The microbial populations participate in the formation of activated sludge, a term commonly used to define a heterogeneous assembly of microorganisms organized in aggregates named "flocs," where bacteria that produce exopolymers promote the attachment of microorganisms and inert matter (Perez et al., 2006; Liu & Fang, 2010). Hence, the actual "heart" of activated sludge systems is the development and maintenance of a mixed microbial culture able to clean the wastewaters, but the culture is sensitive and manipulable as it depends on the organic load, inorganic pollutants, temperature, and nature of industrial discharges (Vigueras-Carmona et al., 2011). This fact can induce proliferation of filamentous bacteria, generating bulking, and make the sludge not to settle correctly, affecting the quality of treated water; some of these bacteria could also form foaming. Both microbial phenomena affect the operation conditions of wastewater treatment plants. Thus, many attempts to thwart bulking and foaming have been made, but most have failed due to a lack of proper scientific foundation to support the efforts of wastewater treatment plants management. Because activated sludge is an essential component of wastewater treatment plants, the final quality of the water depends on the microbial composition of the activated sludge, so the identification and quantification of specific bacteria is relevant, especially when undesirable bacteria proliferate in the flocs, causing operational problems like bulking. In view of this need, different methods have been developed to identify and quantify bacteria in the activated sludge, which have specific limitations. Among them is optical microscopy, which is indispensable to obtain information about the actual quality of the flocs of the sludge, allowing to follow the treatment process accurately and to detect possible deteriorations of the sludge quality before operative problems are generated. An optical microscope is required, next to simple techniques of bright-field and phase-contrasts, as well as techniques of Gram and Ziehl-Neelsen staining for the identification of filamentous microorganisms. Eikelboom and van Buijsen (1981) published a specific microscopy manual for the identification of filamentous organisms; they showed how to investigate microscopically an activated sludge, presenting a compendium of identified filaments in mixed liquor and foam samples from treatment plants under conditions of bulking and non-bulking. Recently, an image analysis method has been developed (using the program ImageJ) to quantify the abundance of protruding filamentous bacteria based on filament length relative to floc area, confirming that filamentous bacteria are necessary to enhance floc stability but if excessively abundant they will adversely affect solid-liquid separation (Dias et al., 2016). However, the identification of filamentous organisms is based mainly on their shapes and sizes, therefore, counting them is focused only on more or less straight filaments, since highly curved filaments form tangles that grow through the floc and cannot be measured in this way, without forgetting that activated sludge in not a mono-culture. In summary, optical microscopy techniques are the fastest method and give a gross estimation of the number of filamentous organisms, but for a large number of sludge samples this method is not very suitable.

Another technical problem is the microbial growth toward the inside of the floc, which is excluded from the optical field of the microscope and limits the identification of filamentous microorganisms. In addition, the composition of inorganic matter influences the microbial development of the flocs. Both facts have generated the need to incorporate the electron microscope to identify the microorganisms inside the activated sludge. The scanning electron microscope (SEM) is an instrument to obtain three-dimensional images with high resolution and great depth of field, in which the ultrastructure of microscopic samples can be observed in great detail, that is, their morphological and topographic characteristics, see Figure 1 (Smoczyn et al., 2014). Also, specimens or organic and inorganic materials can be observed. On the other hand, an instrument that has been developed for the analysis of samples through the SEM is the X-ray energy dispersion spectrometer (EDS) analyzer, which identifies the quantitative and qualitative distribution of the chemical elements that are present in the sample, showing graphs and images related to their distribution; the X-ray EDS analyzer identifies and evaluates the content of chemical elements, from carbon to uranium, on surfaces or flat sections. Therefore, the electron microscopic scanner and the X-ray EDS analyzer have been relevant in the identification of filaments and in the detection and distribution of toxic materials such as heavy metals. Knowledge of the toxicity of heavy metals in growing microorganisms has been widely studied, the presence of metals in the wastewater can damage or destroy the biomass of the flocs, causing deflocculating because its structure cannot be maintained. A turbid effluent appears since settleability in the secondary clarifier is affected. On the other hand, a high load of organic material can be synonymous of abundance and diversity of substrates, favoring the growth of microbial populations, including undesirable organisms such as filamentous bacteria that cause bulking of flocs and foam formation. Thus, electron microscopy generates more information on the state of activated sledges.

Traditional microbial isolation and cultivation techniques have been used to identify and quantify filamentous bacteria in activated sludges. However, these methods can isolate only a portion of bacterial strains, generally described and classified by their phenotype. The phenotype is a very broad term that encompasses the observable traits of the cell, such as the morphology, physiological activity, structure of the cell components, and, in some cases, the ecological niche that the cell occupies. Therefore, microbial methods require a great knowledge and experience in culture media and microbial description, also requires a large consumption of laboratory supplies and time. Fortunately, molecular biology is a big ally for further identifying filamentous bacteria, that is, it allows determining phylogenetic relationships of microorganims, identifying the microbial species, and establishing their fisiologycal characteristics. Besides, it allows determining which microbial populations are active metabolically, as well as monitoring cultivation-independent phylogenetically-defined populations in the activated sludge, leading to a dynamic study of the wastewater depuration process. Therefore, the molecular tools are a new approach to analyze activated sludges, by characterizing the DNA or RNA from a sample without cultivation procedures (Theron & Cloete, 2000) This approach has been successfully applied on soils, clays, sediments, and biofilm.

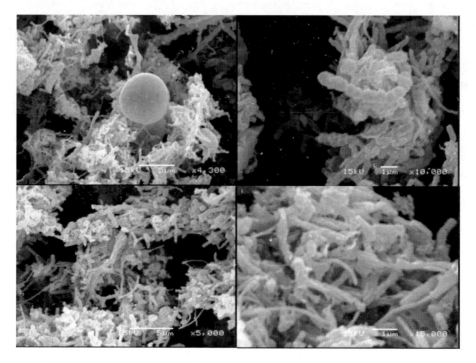

Figure 1. Micrographs of a sample of activated sludge. The images were taken with SEM at different scales. In the images, filamentous microorganisms are observed.

METHODS

Molecular biology technology offers methodologies for the research of environmental samples free of microbial culture, which limits the knowledge of the microbial populations present in activated sludge's (Asvapathanagul et al., 2010; Theron & Cloete, 2000). Techniques such as PCR, PCR-cloning and clone library analysis, FISH with rRNA-targeted nucleic acid probes, qPCR, pyrosequencing combined with phylogenetic analysis are performed to identify detailed changes of bacterial and filamentous bacterial communities in activated sludges.

Polymerase Chain Reaction (PCR)

The polymerase chain reaction was developed by Dr. Kary Mullis in the early 80's (Mullis, 1990); he shared the Nobel Prize in chemistry with Michael Smith in 1993. In the book "Doing PCR: a history of biotechnology," Paul Rabinow tells in detail the historical events that led to the invention of the technique.

Polymerase chain reaction (PCR) is a powerful method for amplifying particular segments of DNA, this reaction is based on the biochemistry of DNA replication, but the procedure is carried out in vitro. PCR uses the enzyme DNA polymerase that directs the synthesis of DNA from deoxynucleotide substrates on a single-stranded DNA template. DNA polymerase adds nucleotides to the 3` end of a custom-designed oligonucleotide when it is annealed to a longer template DNA. Thus, if an oligonucleotide is annealed to a single-stranded template that contains a region complementary to the oligonucleotide, DNA polymerase can use the oligonucleotide as a primer and elongate its 3` end to generate an extended region of double stranded DNA. The PCR is carried out under temperature changes that consider three fundamental steps. First step, the DNA is denatured at more than 90 °C, it produces single-stranded DNA chains, this is named *denaturation*. Second step, the temperature is decreased and the oligonucleotides bind (anneal) to their complementary sequence in the template $_{s-s}$DNA, the oligonucleotides will serve as primers to synthetize DNA. This step is known as *annealing*, the temperature of alignment is calculated by the equation:

$$T_m = 4(G+C)+2(A+T)$$

where T_m is defineted as the temperature at which 50% of the molecules are denatured, and it can be considered as a first aproximation. Third step, named *extension*, the temperature is increased to the polymerization temperature of DNA-polymerase. *Taq* DNA polymerase is the enzyme most used and acts at 72°C, it is a thermostable enzyme isolated from *Thermus aquaticus* (*Taq*), a thermophilic bacterium (50-80 °C) (Sadeghi et al., 2006). The enzyme extends the primers, adding nucleotides onto the primer in a sequential manner, using the target DNA as a template. The reagents required for PCR are DNA template, $MgCl_2$ or $Mg(SO_4)_2$, primers, deoxinucleotide triphosphate (dNTP), enzyme-buffer, and DNA polymerase; sometimes, BSA is used to remove impurities, if the DNA template is rich in GC, formamide or DMSO can be added to facilitate the denaturation of DNA. Finaly, these steps form a cicle that is repeated a number of times and more copies of DNA are generated, and the number of copies of the template is increased exponentially, see Figure 2. Therefore, the synthesis of DNA is done in a specific equipment called thermocycler, of which there are different models with a versatility of functions, depending on the needs of the research. At the end of the PCR, the raction product or amplicon is visualized through an electrophoresis in agarose or polyacrylamide gels, acording to the size of the obtaned amplicons.

Among the techniques of modern molecular biology, PCR is undoubtedly the "queen" technique, since other variants of the PCR have been developed from it, such as real-time PCR (rt-PCR), quantitative real time PCR (rt-qPCR), reverse transcriptase PCR (RT-PCR), multiplex PCR, nested PCR, long-range PCR, fast-cycling PCR, hot start PCR, high-fidelity PCR, etc. In the case of rt-PCR, this allows evaluating the reaction of DNA synthesis in real time, and if the PCR uses reverse transcriptase, the DNA is synthesized from mRNA. The reverse transcriptase enzyme is able to convert the mRNA into a complemetary DNA (cDNA) molecule; and cDNA is used to analyze the expression of the mRNA of a gene of interest. This method was copied from retroviruses that use a reverse transcriptase to convert their RNA genome into DNA duplicated in millions of viral particles. The quantitative term refers to quantify the amount of DNA in the sample; if the reaction uses genomic DNA then it is a qPCR, if cDNA by reverse trancriptase is obtained first and then the PCR is done, this refers to an RT-qPCR. Two great advantages of rt-PCR are highlighted, the amplification product is quantified during the reaction and an agarose gel is not nesesary to know if the reaction was successful, as in the endpoint PCR.

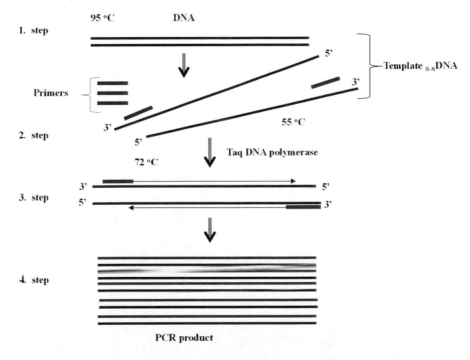

Figure 2. Polymerase chain reaction (PCR). 1. Step: Denaturation, DNA-target is separated into simple chains, cDNA. 2. Step: Annealing, the primer sequence hybridizes to complementary regions on cDNA. 3. Step: Extension, DNA synthesis produces long-templates from the original strands. The three steps form a cycle, and this is repeated up to 35 times. 4. Step: Final extension, PCR product.

PCR and Cloning

In the process of cloning, the target DNA molecules, which are the object of study, bind to autonomous DNA such as plasmids or viruses (vectors), generating hybrid molecules called recombinant DNA or recombinant vectors, which are introduced into a host cell, in which the recombinant vector is replicated and recovered for the study. The cloned DNA can be characterized by techniques such as restriction mapping, expression induction, and nucleotide sequence, among others. Likewise, a typical cloning experiment requires of the construction of a recombinant DNA library and a detection system for clones, like the selection of white or blue clones, and an isolation method of recombinat vectors. Thus, a library is the set of DNA fragments of an organism distributed in vectors, whose binding represents the total genome of the organism. In practice, a library contains some or all the genes of a given species (Figure 3). In this context, the PCR has facilitated gene cloning or of specific fragments of DNA, therefore, a library is constructed of DNA or cDNA from mRNA.

In particular, the construction of libraries followed by a Sanger sequencing of small subunit rRNA gene (16S rRNA gene in Bacteria and Archaea or 18S rRNA gene in Eukarya) have been widely applied to study the composition, organization, and space-time patterns of microbial communities (Burrell et al., 1998, Woese, 1987). The libraries based on 16S rRNA genes have been used to characterize phylogenetic structure of microbial communities in environmental samples. Phylogenetic study of a microbial community provides the phylogenetic affiliation of the individual populations and gives an approximate estimate of the diversity of the predominant species. The concept of microbial diversity has been defined as the range of significantly different types of organisms and their relative abundance in an assemblage or community. The diversity has also been defined according to information theory, as the amount and distribution of information in an assemblage or community (Torsvik et al., 1998). Microbial diversity refers unequivocally to biological diversity at three levels: within species (genetic), species number (species), and community (ecological) diversity (Harpole, 2010).

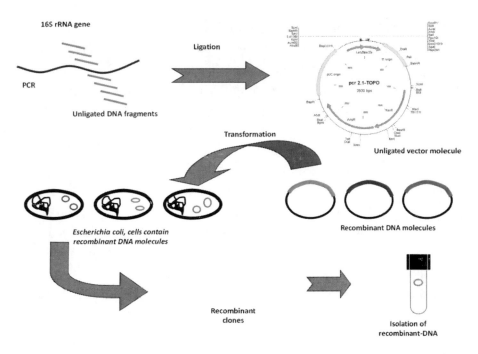

Figure 3. Global scheme of the cloning process. TOPO vector cloning (Thermo Fisher Scientific, US).

In wastewater treatment, the activated sludge process is extensively used due its high microbial diversity and activity, resulting in the removal of most organic pollutants and nutrients (Perez et al., 2006). The composition and diversity of the microbial community has the greatest impact on stability and performance of the wastewater treatment systems (Lui & Fang, 2010). A biological community of activated sludge has a large variety of viruses, bacteria, protozoa, fungi, algae, and metazoan, where the bacteria are the most abundant organisms (95%) and play a crucial part in wastewater treatment. However, bulking sludge due to overgrowth of filamentous bacteria and/or Zoogloea organisms have a strong influence on the performance of the activated sludge system; the overgrowth of filamentous bacteria in the activated sludge affects the operation of the wastewater treatment plant, as mentioned above (Xu et al., 2018).

16S rRNA gene clone libraries have been constructed from activated sludge samples, the results of phylogenetic analysis of the obtained sequences have shown that Proteobacteria are the most abundant phylogenetic group, followed by *Bacteroidetes* and *Firmicutes*. Some specific genera, e.g., *Nitrosomonas*, *Thauera*, and *Dechloromonas*, significantly correlate with the functions and performance of wastewater treatment. Also, a large number of unclassified sequences are reported in the library, suggesting that a wide variety of novel species may inhabit complex activated sludge communities. The structure of the bacterial community did not differ significantly among samples of sludge analyzed in this way. Further, the diversity of the microbial communities suggests that activated sludges possess high metabolic potential and equivalent functions as required for wastewater treatment. On the other side, the dominance of *Proteobacteria* has also been demonstrated by PCR-denaturing gradient gel electrophoresis (PCR-DGGE) and terminal restriction fragment length polymorphism (T-RFLP). DGGE is used to investigate microbial communities through detecting the information of nucleic acid and can allow the observation of uncultivable bacteria and their functional genes (Muyzer et al., 1998; Reddy et al., 2016). Hence, some specific genera, e.g., *Nitrosomonas*, *Thauera*, and *Dechloromonas* correlate significantly with the functions and performance of wastewater treatment, and are abundant in activated sludges (Shah, 2016).

Quantitative Real Time PCR

Knowledge on the diversity of microbial communities, richness of species, and the determination of the dominant species in the activated sludge have allowed to determine the metabolic activity of the populations based on their phylogenetic relationships. However, 90% of the microbial population is not cultivable, which leads us to explore the metabolic activity through other molecular ways that will generate more accurate information during the wastewater treatment process. An alternative molecular pathway is quantitative real-time PCR. The rt-qPCR generates information on the metabolism carried out within the activated sludge, and which are the active metabolic populations. This technique is also used for dynamic studies on wastewater depuration processes, and studies on the dynamics of populations can be carried out in terms of the metabolic function in real time (Livak & Schmittgen, 2001).

Real-time PCR is the most sensitive method to detect and quantify nucleic acids. In principle, the technique is based on the theoretical fundamentals of end point PCR, and one of the applications most used is quantification of the gene expression, even at very low levels of mRNA expression in cells. For rt-qPCR amplification, it is necessary to use a fluorophore to determine the number of mRNA molecules, previously a reverse transcription reaction must be performed to form cDNA; and then, amplification of the DNA is performed in a thermal cycler coupled to an optical system, which monitors the signal of the fluorophores used to detect the amplified product.Monitoring of the products amplified during the reaction is done by the use of fluorescent reporters, these follow two different methods: specific and non-specific. Non-specific methods are based on the use of intercalating molecules that have an affinity for double-stranded DNA and their oxidizer generates a fluorescent signal, which is captured in the extension stage of each cycle and is proportional to the number of double-stranded DNA copies obtained in each cycle of the PCR. The most used reporter is called SYBR Green because it is practical and inexpensive, the molecule has a positive charge, and while it is in solution the binding to double-stranded DNA is very low and practically does not emit fluorescence. However, when it binds to the smaller groove of the DNA, it increases its fluorescence up to 1000 times. The great disadvantage of SYBER Green is that it can bind to any double-stranded DNA molecule, including primer dimers. This methodology is possible because it follows the principle known as fluorescent resonance energy transfer to generate the signal; the method involves transferring energy from a donor or fluorescent reporter to an acceptor or quencher. Those are probes based on hydrolysis and hybridization. The oligonucleotide probes are labeled with a fluorescent reporter and a quencher, both are closely linked, because the probe does not hybridize to its target sequence, it does not emit fluorescence. When the probe hybridizes, conformational changes occur in the reporter and the quencher, allowing the 5'-3' exonuclease activity of the Taq DNA polymerase to break the binding, the fluorescence is emitted when the reporter is released and captured by the equipment. These methods are very safe, if there is neither union of the probe to the target sequence nor amplification, and no fluorescence occurs; therefore, the specificity is very high. An example of these systems are the commercial probes known as TaqMan (Vanysacker et al., 2014), but others exist in the market.

Gene quantification requires that the thermocyclers are provided with a computer and software capable of generating a series of graphs, which show the necessary data to ascertain that the reaction was successful. One graph is related to the amplification and the progress of the reaction, another graph shows the dissociation curve, providing information about the specificity of the reaction. Another important point of the analysis is the type of quantification that will be used to determine the precise amplification of the gene target; this depends on the interests of the researcher. There are two types of quantification: absolute and relative. The first is used to know the exact number of amplified copies of the gene target or the precise concentration of nucleic acids in a sample. In practice, this type of quantification is used to measure viral or bacterial load in the sample. The second evaluates the changes in the expression of genes in different physiological states. These changes are based on the levels of the mRNA of the target gene compared to a reference gene (housekeeping gene) that does not change its expression even though the physiological states are modified for diverse reasons. Data are expressed as relative to the reference gene and are generally referred to as the number of times in which the mRNA levels increase or decrease. Whatever the type of quantification, the software of the equipment is designed to carry out the mathematical and statistical analysis for each type of quantification. Most thermocyclers have a fluorometric system that consists of source energy to excite the reporters. The source of energy can be of three types: light bulbs, light emitting diodes (LEDs), and lasers, whichever it is, first the reporter is excited and its emission signal is collected through a filter, which allows the passage of the wavelength corresponding to a photodetector that captures the information coming from the sample, and this is analyzed through the software of the equipment, see Figure 4. Other characteristics are the speeds to increase or decrease the temperatures in each stage of the reaction, the number of samples that can support and the consumables for the reaction. Among the devices that use a lamp as a source of excitation, are the ABI Prism 7000 or the 7500 model from Applied Biosystems, the Mx4000 and Mx3000P models from Stratagene and the iCycler iQ model from Bio-Rad. The equipment that uses LEDs is the LightCycler of Roche, SmartCycler of Cepheid, Rotor-Gene of Corbett, and the DNA Engine Opticon of MJ Research. The ABI Prism model 7900HT from Applied Biosystems uses lasers (Wong & Medrano, 2005).

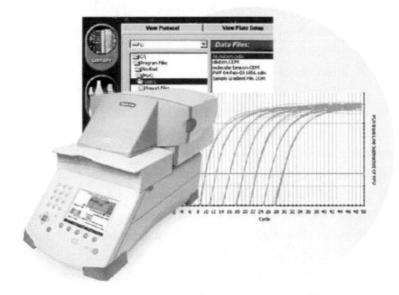

Figure 4. The new MyiQ real-time PCR detection system offers an affordable alternative for the detection of common green fluorescent dyes such as FAM and SYBR Green I. (As of November 27, 2018, Bio-rad listed on its website https://www.gene-quantification.de/platform1.html).

Fluorescence *In Situ* Hybridization (FISH)

From the structural model of the DNA proposed by Watson and Crick, it is well known that it has two chains running in opposite directions forming a double helix, which is joined by hydrogen bonds through the complementarity of nitrogenous bases (purine-pyrimidine): adenine-thymine and guanine-cytosine, each pair is linked by two and three unions, respectively, see Figure 5 (Watson & CricK, 1953). In addition, the duplex DNA is sensitive to increases of temperature and the molecule becomes denatured, generating two single-stranded DNA. When temperature decreases, both DNA strands bind in a complementary manner to form double-stranded DNA or duplex DNA. This biochemical principle of DNA is taken to define nucleic acid hybridization. In the 1960s, molecular hybridization with labeled DNA or RNA was used to identify or quantify the position of DNA sequences *in situ* in a biological sample (Schildkraut et al., 1961). Joseph Gall and Mary Lou Pardue published, that radioactive copies of a ribosomal DNA sequence could be used to detect complementary DNA sequences in the nucleus of a frog egg (Gall & Pardue, 1969). This could have been the starting point of hybridization as a molecular tool of great versatility and sensitivity in cytogenetic. Thus, different hybridization techniques start from two nucleic acid molecules; a homogeneous known sequence that acts as a probe and the other heterogeneous unknown sequence containing the target sequence to be detected. The probe must be region-specific and complementary to the target sequence. Hybridization can be carried out in liquid medium or on a solid support, such as nitrocellulose, where one nucleic acid molecule is bound. The result of the hybridization can be visualized, since one of the chains is marked with a radioactive compound or fluorochrome. If the probe is labeled, the hybridization is standard, but if the labeling is on the target molecule, the hybridization is reversed. In summary, hybridization is based on the union of two simple chains of nucleic acids that gives rise to a double-stranded structure, this can be DNA-DNA, RNA-RNA (both homoduplexes) or DNA-RNA (heteroduplex) hybrids, and the mating takes place by the complementarity of bases through the hydrogen bridges. Therefore, several molecular techniques are based on hybridization, including PCR, northern blot, Southern blot, DNA microarrays, screening of libraries, or in situ hybridization.

In the early 1980s, fluorescence in situ hybridization (FISH) was developed; it uses fluorescent probes, which bind to specific sites of the chromosome with a high degree of sequence complementarity to the probes (Bishop, 2010). The technique allows knowing where and when a specific DNA sequence exists in cells by detecting the fluorescent group. Further, FISH is complemented with other techniques like fluorescence microscopy to find out where the fluorescent probe is bound on the chromosome, and flow cytometry to detect the binding quantitatively. This technique allows a rapid and specific identification of microbial cells whether as individual cells or as grouped cells in their natural environment (Moter & Göbel, 2010). Knowledge of the composition and distribution of microorganisms in natural habitats provides a solid support to understand interactions between different species in the microhabitat. To carry out this type of microbiological studies, the most used target molecules for FISH are ribosomal genes, because they can be found in all living organisms; Bacteria and Archaebacteria contain rRNA genes of 5S, 16S, and 23S, and 5.8S, 18S and 26S rRNA in Eukarya, Figure 6 (Woese, 1987). However, the 16S rRNA gene is the target molecule most popular because it is relatively stable and has a high number of copies; besides, this gene is much conserved and contains highly variable regions. The popularity in microbial phylogenetic studies has generated an increase in sequences of the 16S rRNA gene in the databases, this fact has facilitated the identification by FISH of most microorganisms, especially when it is required to identify microbial populations that cannot be cultivated. The high number of copies of the 16S rRNA gene in each replication and in metabolically active cells offer enough target molecules that allow the visualization of individual bacterial cells, even when they are part of a tissue.

The probes designed using the information of the complementary sequence are end-labeled at the 5'-end. Among the markers most commonly used in FISH for microbiological studies are the fluorescein derivatives, fluorescein isothiocyanate (FITC) and 5-(6) carboxyfluorescein-N-hydroxysuccinimide ester (FluoX).

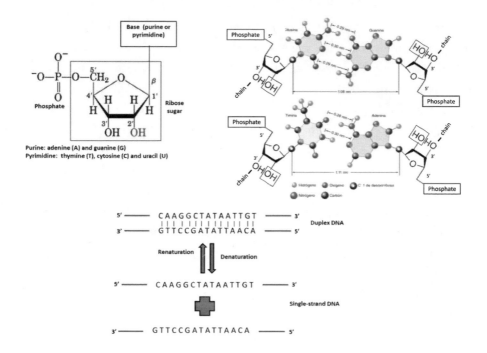

Figure 5. Chemical composition and molecular structure of DNA. Denaturation of DNA by temperature.

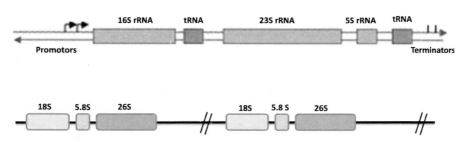

Figure 6. The graphs represent the operons for the ribosomal genes of prokaryotes and eukaryotes. The first operon corresponds to *Escherichia coli* and the next *Saccharomyces cerevisiae*.

The derivatives of rhodamine include tetramethyl rhodamine isothiocyanate (TRITC) and Texas red. Cyanine dyes of wide wavelength are Cy5.5- Cy7, which are excited in the red region of the spectrum (675 nm) and emit in the far red (760 nm). Other fluorophores widely used are of the Alexa fluor family and nanometric glass particles called "quantum dots," these allow to observe an intense brightness and are more photostable (Bishop, 2010). When the selection of a specific site is on rRNA genes, the design of the probe, as well as the type of labeling, should be done with special care because the quality of the hybridization and the visualization of cells depend in part on these features. Cyanine dyes-labeled probes are used in experiments involving conventional flow cytometry and 6-carboxyfluorescein-labeled probes are used for experiments involving microscopy or imaging cytometry.

On the other hand, there are different ways of marking for probes but can be summarized in direct and indirect. Direct fluorescence labeling is the most common, quick, easy, and economical, since it does not require any additional step for detection after hybridization. One or more molecules labeled with the fluorescent compound are directly linked to the oligonucleotide, either chemically during synthesis through amino bonds to the 5 'end of the probe or enzymatically using terminal transferases that bind to the 3' end of the nucleotide. In indirect detection, the sensitivity of FISH is increased by joining the probe to molecules, such as digoxigenin (DIG), and then it is detected by a fluorescent antibody (Figure 7). Also, the use of enzymes amplifies the signal and the sensitivity of FISH. In this case, the oligonucleotides are labeled with horseradish peroxidase (HRP) using fluorescein tyramide (TSA) as substrate. The sensitivity of the technique can be increased by combining polyribonucleotide probes labeled with digoxigenin and TSA, since it has been found that TSA increases between 10 and 20 times the intensity of the signal.

FISH has demonstrated great potential for the analysis of bacterial community composition in different environments. It is commonly analyzed by rRNA-targeted nucleic acid probes, due to probes allow the visualization and quantification of the bacterium behind an rRNA sequence; and rRNA sequences can be retrieved directly from the environment without prior cultivation of the organism of interest. Besides, the identification and quantification of individual cells is done by direct microscopic observation (Hug et al., 2005).

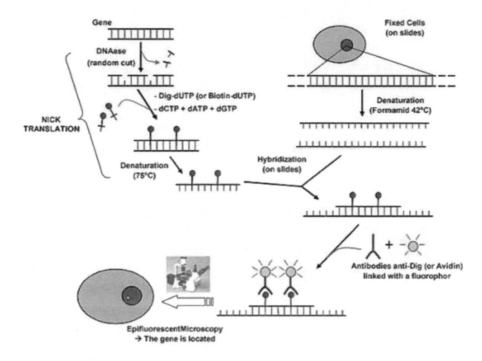

Figure 7. The schematic diagram of fluorescence in situ hybridization (FISH). The resource comes from Wikipedia (As of November 26, 2018, Wikipedia on its website https://www.creativebiomart.net/resource/principle-protocol-fluorescence-in-situ-hybridization-fish-protocol-342.htm)

This method is suitable for samples that are difficult to be processed by semi-automated image analysis techniques, such as activated sludge's. As already mentioned, an activated sludge is a complex and highly variable mixture of different microorganisms, which are fundamental for the cleaning of wastewater. Monitoring population changes of the organisms permits the identification of bacteria that induce bulking and foam formation in a temporal manner in activated sludge's. This is possibly produced by the organic matter load of the effluent, since the concentration and type of substrates can vary, causing changes in the population dynamics and in the metabolism of organisms. Hung and collaborators proposed a quantification protocol by FISH that allows hybridizing the samples on slides and their analysis by direct microscopic observation (Hung et al., 2005). The abundance categories were designed based on digital images of the target organisms for the filamentous bacterium *Microthrix parvicella* and for different morphotypes of nocardioform actinomycetes, but can easily be adapted to other types of microorganisms. This protocol must be rapid, possible with an epifluorescence microscope, no need of an expensive confocal laser-scanning microscope, suitable for a difficult matrix containing debris and cells of different fluorescence intensity, appropriate for FISH protocols requiring cell wall permeabilization, allowing for quantitative comparison between different concentrations of one organism, as well as between different target organisms, it is of small uncertainty due to subjectivity of operators, accurate enough to identify typical seasonal changes in the population composition.

The fluorescent in situ hybridization (FISH) and other 16S rRNA-based techniques were used to study bulking bacteria, they revealed many unclassified groups. To our knowledge, there are more than 20 types and over 30 cultured species of foaming bacteria. Molecular methods, such as FISH or rt-qPCR, are highly sensitive and standardized; these methods are limited by specific probes or primers and could not give an overall profile of the bulking and foaming bacteria. Other techniques such as PCR-cloning and clone library analysis or pyrosequencing combined with phylogenetical analysis are performed to identify detailed changes of bacterial and filamentous bacterial communities in activated sludges.

High-Throughput Sequencing

In 1977, Sanger published a sequencing method in the enzymatic synthesis of DNA chains using DNA polymerase and a specific terminator. The reaction generates DNA fragments with different sizes and easily identifiable by a specific terminator (nucleotides that do not have a hydroxyl group at its 3 'end (ddNTP)). The fragments can be separated into an electrophoresis gel with four different lanes to determine the sequence of the DNA template (Sanger et al., 1977). But years later, nucleotides labeled fluorescently were incorporated to the sequencing of

Sanger (Smith et al., 1986; Lee et al., 1992). In time, the Sanger´s method was taken as the basis to develop automatized sequencing. However, the need of obtaining the most information and faster with cheaper methods, new sequencing technologies have been developed, thus, the high throughput sequencing has replaced the elegant Sanger method. High-throughput sequencing, also called "new generation" or "massive sequencing," is a method used to sequence thousands of millions of different DNA fragments at the same time (Rodríguez-Santiago & Armegol, 2012). For example: sequence of Sanger provides a definitive resolution to detect small genetic variants but has the limitation of only being able to perform 96 or 384 reactions in parallel. This causes prolonged sequencing time and the price per sequenced base is high. In consequence, the cost for ambitious projects is estimated in US$ 0.5 per kilobase (kb), or approximately €1 for each 2.5 kb (Metzker, 2005; Tucker, 2009). Therefore, the new platforms of sequencing have the capacity to sequence millions of DNA fragments in parallel at much cheaper price per base (Rodríguez-Santiago and Armengol; 2012). In addition, high-throughput sequencing has the potential to detect all types of genomic variations in a single experiment. This new generation of sequencing has forged the way for the discovery of new genes, proteins, and biochemical pathways, among other scientific achievements. Introduction of the high throughput sequencing offers new ways to study the microbial communities of many environmental sample types, such as activated sludge, without the need to carry out the cloning of ribosomal genes or specific genes for a metabolic function.

The sequencing technologies implemented in the different instruments currently used for the high-throughput sequencing differ in several aspects, but the main working scheme is conceptually similar for all of them (Metzker, 2010; Voelkerding et al., 2009). As for all sequencing, a DNA polymerase and nucleotides are needed, but the DNA must be fragmented previously, then the fragments must be linked to adapter sequences by their ends. The DNA fragments are amplified and clustered to be sequenced. These fragments are adhered to a surface that can be nanospheres or plates, with the purpose of facilitating their manipulation and analysis (Tucker et al., 2009). High-throughput sequencing by plates uses labeled nucleotides that are added to the DNA attached to a plate. A device takes pictures of the plate of all the nucleotides incorporated in the multiple sequences. In the case of nanospheres, sequencing is carried out by means of two reactions: the nucleotides are first incorporated one by one (A, T, C, and G). These react with a protein and emit a fluorescent light, indicating the nucleotide that has been incorporated into the DNA template. Subsequently, this light is detected and stored in a computer file, thus determining the nucleotide sequence of multiple DNA fragments (Alquicira, 2016).

The sequencing platforms can produce more than 35 million sequences of 400 bp for each run, providing detailed information on bacterial community structure qualitatively and quantitatively. Assuming all bacteria in a sludge have the same copy number of the 16S rRNA gene in their genomes, the detection limit is about 0.01% if 10,000 reads were obtained for the sample. Thus, it could provide much more information than other 16S rRNA-based or microscopic methods. To date, the cost and operational time of high throughput sequencing is reduced continuously. It is a promising way to monitor bacterial communities in activated sludges in an automated and precise manner. Moreover, providing a theoretical basis for developing specific strategies to control sludge bulking and other operational problems of wastewater treatment plants, already described. These new technologies offer a particularly valuable use for metagenomics, since they can be the basis for metranscriptomics and metaproteomics of activated sludges.

For the study of microbial communities whether by PCR-cloning or high-throughput sequencing, or another way, it is necessary to recover a metagenomic DNA of high quality and purity. Fortunately, a large number of methods have been published to obtain metagenomic DNA from different environmental samples, including activated sludge samples. Hence, the researcher can look for the best method that provides the quality of DNA required for the technique or molecular techniques to be developed, or if having ample experience in the field of molecular biology, the investigator can design the most appropriate method for the needs of the research.

Particularly, studies to analyze the bacterial community compositions by high-throughput sequencing have been reported, aimed at identifying the bacteria that cause bulking of activated sludge sludge's. Activated sludge samples were collected from different wastewater treatment plants in the northern Xinjiang Uygur Autonomous Region of China in winter (Xu et al., 2018). The composition of the microbial community was analyzed through the amplification of the variable region V1-V5 of the rRNA 16S gene from metagenomic DNA; the amplicons were sequenced by high-throughput sequencing. The sequencing generated 30087–55170 effective reads, representing 36 phyla, 293 families, and 579 genera in all samples. In them, Proteobacteria was the most dominant phylum (26.7–48.9%) in all samples. Bacteroidetes, Chlorofexi, and Actinobacteria were the other important groups, comprising 19.3–37.3%, 2.9–17.1%, and 1.5–13.8%, respectively. In addition, several phyla accounted for more than 1% in at least one sample, for example, Firmicutes (1.4–4.8%), Planctomycetes (1.7–4.5%), Chlorobi (0.2–3.8%), Acidobacteria (0.5–5.1%), Saccharibacteria (0.4–3.2%), and Ignavibacteria (0.03–1.3%). Among Proteobacteria, Betaproteobacteria was the most abundant class (34.4–65.8%), 11 orders were identified of which Burkholderiales was the predominant main order, between 44.1 and 82.5% of all samples. Methylophilales, Neisseriales, Nitrosomonadales, and Procabacteriales were detected with lower abundances. Another similar work has reported that the genera *Actinobacteria*, *Clostridium* XI, *Arcobacter*, *Flavo-bacterium* have been more abundant in the foam than in activated sludges, undescribed species have also been identified, such as *Gordonia* (Gou et al., 2015; Gou & Zhang, 2012). However, the bacterial populations can change with respect to the conditions of the effluent, season of the year, and temperature (Burger et al., 2017, Shchegolkova et al., 2016, Wang et al., 2016).

REFERENCES

Alquicira, José. 2016. *"High performance sequencing (Pyrosequencing and Illumina)."* December 10, 2018. Conogasi.org. Website: http://conogasi.org/articulos/secuenciacion-de-alto-rendimiento-pirosecuenciacion-e-illumina/.

Asvapathanagul, Pitiporn, Hyeeun Bang, Hyeyoung Lee, Betty H. Olson. 2010. "Concurrent Rapid Identification of Bulking and Foaming Bacteria." *Water Environment Federation* 2010: 587-600.

Bio-Rad. 2018. Cycler.gene.quantyfication.info. Accessed November 27, 2018. (https://www.gene-quantification.de/platform1.html)/ de León Gallegos, Martin Denecke, Philipp Wiedemann, Fabio K. Schneider, Hajo Suhr. 2016. *"Image processing for identification and quantification of filamentous bacteria in situ acquired images."* 15:64. doi 10.1186/s12938-016-0197-7.

Bishop, Ryan. 2010. "Applications of fluorescence in situ hybridization (FISH) in detecting genetic aberrations of medical significance." *Bioscience Horizons* 3: 86-95. Accessed March, 2010. http://creative commons.org/licenses/by-nc/2.5.

Burger, W., K., P. J. Scales Krysiak-Baltyn, G. J. O. Martin, A. D. Stickland, S. L. Gras. 2017. "The influence of protruding filamentous bacteria on floc stability and solid-liquid separation in the activated sludge process." *Water Research* 123: 578-585.

Burrell, Paul C., Heather Gwilliam, Debbie Bradford, Philip L. Bond, Philip Linda L. Blackall Hugenholtz. 1998. "The use of 16S rDNA clone libraries to describe the microbial diversity of activated sludge communities." *Water Science and Technology* 37: 451-454.

Eikelboom, Ir. D. H. and H. J. J. van Buijsen. 1981. *"Microscopic sludge investigation manual."* The Netherlands. TNO Research Institute for Environmental Hygiene. (Eikelboom and van Buijsen 1983, 29-34).

Gall, J. G., and Pardue, M. L. 1969. "Formation and detection of RNA-DNA hybrid molecules in cytological preparations." *Proceedings of the National Academy of Sciences* 63, 378-383.

Guo, Feng, Zhi-Ping Wang, Ke Yu, T. Zhang. 2015. "Detailed investigation of the microbial community in foaming activated sludge reveals novel foam formers." *Scientific Reports* 7637 (2015). Accessed January 6, 2015. https://www.nature.com/articles/srep 07637.

Guo, Feng, T. Zhang. 2012. "Profiling bulking and foaming bacteria in activated sludge by high throughput sequencing." *Water Research* 46: 2772-2782.

Hug, Thomas, Willi Gujer, Hansruedi Siegrist. 2005. "Rapid quantification of bacteria in activated sludge using fluorescence in situ hybridization and epifluorescence microscopy." *Water Research* 39: 3837-3848.

Liu, Yan, Herbert H. P. Fang. 2010. "Influences of Extracellular Polymeric Substances (EPS) on Flocculation, Settling, and Dewatering of Activated Sludge." *Critical Reviews in Environmental Science and Technology* 33: 237-273.

Livak K. J. and T. D. Schmittgen. 2001. "Analysis of relative gene expression data using real-time quantitative PCR and the 2(-Delta Delta C (T)) method." *Methods* 25: 402-408.

Mesquita, D. P., A. L. Amaral, E. C. Ferreira. 2011. "Identifying different types of bulking in an activated sludge system through quantitative image analysis." *Chemosphere* 85: 643-652.

Metzker ML. 2010. "Sequencing technologies - the next generation." *Nature Reviews Genetics* 11:31-46.

Metzker, ML. 2005. "Emerging technologies in DNA sequencing." *Genome research* 15:1767-76. 3.

Moter A. and U. B. Göbel. 2010. "Fluorescence in situ hybridization (FISH) for direct visualization of microorganisms." *Journal Microbioloy Methods* 41: 85-112.

Mullis Kary B. 1990. "The unusual origin of the polymerase chain reaction." *Scientific American* 262: 56-61.

Muyzer, Gerard and Kornelia Smalla. 1998. "Application of denaturing gradient gel electrophoresis (DGGE) and temperature gradient gel electrophoresis (TGGE) in microbial ecology." *Antonie Leeuwenhoek* 73: 127-141.

Pardue, Jary Lou and Joseph G. Gall. 1969. "Molecular hybridization of radioactive DNA to the DNA of cytological preparations." *Genetics* 64: 600-604.

Perez, Y. G., S. G. F. Leite, M. A. Z. Coelho. 2006. "Activated sludge morphology characterization through an image analysis procedure." *Brazilian Journal of Chemical Engineering* 23: 319-330.

Reddy, GV, Hiral Borasiya, Shah MP. 2016. "Determination and Characterization of Microbial Community Structure of Activated Sludge." *Advances in Recycling & Waste Management* 1:2. Accessed June 28, 2016. doi: 10.4172/2475-7675.1000110.

Rodríguez-Santiago, Benjamín, Lluís Armengol. 2012. "Tecnologías de secuenciación de nueva generación en diagnóstico genético pre- y postnatal." *Diagnóstico Prenatal* 23: 56-66. Accessed April 6, 2012, Pages. doi:10.1016/j.diapre.2012.02.001. [Next generation sequencing technology in pre- and postnatal genetic diagnosis. *Diagn Prenat 23*: 56-66.]

Sadeghi, H. Mir Mohammad, M. Rabbani, F. Moazen. 2006. "Amplification and cloning of Taq DNA polymerase gene from Thermus Aquaticus YT-1 *Research in Pharmaceutical Sciences* 1: 49-52.

Sanger, Frederick, S. Nicklen, A. R. Coulson. 1977. "DNA sequencing with chain terminating inhibitors." *Proceedings of the National Academy of Sciences* 74: 5463-5467.

Schildkraut, Carl L, Julius Marmur, Paul Doty. 1961. "The formation of hybrid DNA molecules and their use in studies of DNA homologies." *Journal of Molecular Biology* 3: 595-617.

Shah, MP. 2016. "Microbial Community Structure of Activated Sludge As Investigated With DGGE." *Journal of Advanced Research in Biotechnology* 1: 2-7. Accessed November 24, 2016. doi: http://dx. doi.org/10.15226/2475-4714/1/1/00111.

Shchegolkova, Nataliya M., George S. Krasnov, Anastasia A. Beloval, Alexey A. Dmitriev, Sergey L. Kharitonov, Kseniya M. Klimina, Nataliya V. Melnikova, Anna V. Kudryavtseva. 2016. "Microbial Community Structure of Activated Sludge in Treatment Plants with

Different Wastewater Compositions." *Frontiers in Microbiology.* Accessed February 18, 2016. doi: 10.3389/fmicb.2016.00090.

Smoczyn, Lech, Harsha Ratnaweera, Marta Kosobucka, Michal Smoczyn. 2014. "Image analysis of sludge aggregates" *Separation and Purification Technology* 112:412-420.

Theron J. and Cloete T. E., 2000. "Molecular techniques for determining microbial diversity and community structure in natural environments" *Critical Reviews in Microbiology* 26: 37-57.

Tucker Tracy, Marco Marra, Juan M Friedman. 2009. "Massively parallel sequencing: the next big thing in genetic medicine." *American Journal of Human Genetics* 85:142-54.

Vanysacker, Louise, Carla Denis, Joris Roels, Kirke Verhaeghe, Ivo F. J. Vankelecom. 2014. "Development and evaluation of a TaqMan duplex real-time PCR quantification method for reliable enumeration of *Candidatus Microthrix.*" *Journal of Microbiological Methods* 97: 6-14.

Vigueras-Carmona, Sergio E., Florina Ramírez, Alberto Noyola and Oscar Monroy. 2011. "Effect of thermal alkaline pretreatment on the anaerobic digestion of wasted activated sludge." *Water Science and Technology* 64: 953-959.

Voelkerding, Karl V., Shale A. Dames, Jacob D. Durtschi, 2009. "Next-Generation Sequencing: From Basic Research to Diagnostics." *Clinical Chemistry* 55: 641-658. Accessed October 7, 2008. doi: 10.1373/clinchem.2008.112789.

Wang, P., Z. Yu, R. Qi, H. Zhang 2016. "Detailed comparison of bacterial communities during seasonal sludge bulking in a municipal wastewater treatment plant." *Water Research* 105:157-166.

Watson, J. D., and F. H. C Crick. 1953. "Molecular structure of nucleic acids: A structure for deoxyribose nucleic acid." *Nature* 171; 737-738. doi:10.1038/171737a0.

Wikipedia. (https://www.creativebiomart.net/resource/principle-protocol-fluorescence-in-situ-hybridization-fish-protocol-342.htm, 26/11/2018).

Woese, Carl R. 1987. "Bacterial evolution." *Microbiological Reviews* 51: 221-271.

Wong, M. L. and J. F. Medrano. 2005. "Real-time PCR for mRNA quantitation." *BioTechniques* 39: 75-85.

Xu, Shuang, Junqin Yao, Meihaguli Ainiwaer, Ying Hong, and Yanjiang Zhang. 2018. "Analysis of bacterial community structure of activated sludge from wastewater treatment plants in winter." *BioMed Research International* 2018, Article ID 8278970, 8 pages. https://doi.org/10.1155/2018/8278970 .

In: The Activated Sludge Process   ISBN: 978-1-53615-202-9
Editor: Benjamin Lefèbvre   © 2019 Nova Science Publishers, Inc.

Chapter 3

A MACROKINETIC USE OF THE MONOD MODEL IN BIOLOGICAL WASTEWATER TREATMENT

Mario Plattes, PhD
Environmental Research and Innovation (ERIN)
Luxembourg Institute of Science and Technology (LIST)
Esch-sur-Alzette, Luxembourg

ABSTRACT

The Monod model gives a functional relation between specific growth rate and substrate concentration in the bulk; the maximum specific growth rate and the half-saturation index being the parameters. State of the art activated sludge models are based on the Monod model and have found wide application in engineering practice. However, the Monod model remains empirical only and a fundamental explanation of the Monod model remains unknown, although it is believed that the values of the Monod parameters are a function of both diffusional mass transport outside and inside the activated sludge floc and enzyme kinetics inside the cells. An explanation of the Monod model might therefore be given by Fick's laws of diffusion and the law of mass action (or

Michaelis-Menten kinetics). The application of Fick's laws of diffusion requires, however, spatial dimensionality, which is a problem since activated sludge flocs have a complex three-dimensional structure, a feature that seems to be elusive to be described adequately by a mathematical model. Note, that the same problem is faced when modelling biofilm systems, and therefore zero-dimensional biofilm models have recently been proposed, overcoming the problem of modelling the complex biofilm structure. The zero-dimensional modelling approach using the Monod model has thus been established in activated sludge modelling and analogously in biofilm modelling, bearing in mind that both systems consist of similar cell aggregates. The Monod model as used in the zero-dimensional approach describes macrokinetic behaviour of biological wastewater treatment systems rather than intrinsic kinetics of activated sludge flocs or biofilms. This comment will briefly review important research efforts dedicated to an adequate use of the Monod model, consolidate knowledge from activated sludge and biofilm modelling, identify misdirections, and set directions for further research towards a unified macrokinetic use of the Monod model in biological wastewater treatment.

OUTLINE

- The Monod model and its use in biological wastewater treatment modelling
- A unified view of growth processes in biological wastewater treatment
- Why think and model biofilms in layers?
- A quality assessment of zero-dimensional models for biological wastewater treatment
- Future directions

THE MONOD MODEL AND ITS USE IN BIOLOGICAL WASTEWATER TREATMENT MODELLING

The Monod model [1] is widely used for modelling microbial growth on a given substrate. The Monod equation (Eq. 1) gives a functional relation between the specific growth rate (μ), maximum specific growth rate (μmax), substrate concentration (S), and Monod affinity constant or half-saturation coefficient (K_S), now called half-saturation index [2]. The functional relation given by the Monod equation is illustrated in figure 1.

$$\mu = \mu_{max} \cdot \frac{S}{K_S + S} \qquad (1)$$

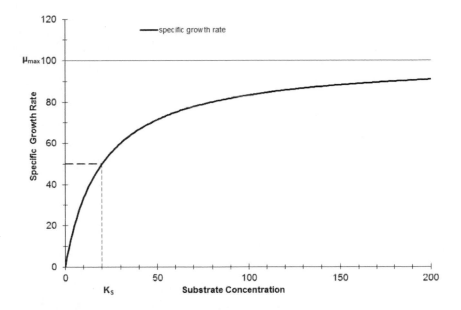

Figure 1. Graphical presentation of the Monod model showing the relation between specific growth rate and substrate concentration (μmax = 100 and K_S = 20).

Although there have been attempts to give a mechanistic explanation of the Monod equation [3, 4] and although it is mathematically the same as the Michaelis-Menten equation [5], the Monod equation remains empirical only.

In state of the art activated sludge models (ASMs) [6], which are zero-dimensional (0D) [7], the Monod model is used to model the macrokinetic behaviour of activated sludge systems, not the intrinsic kinetics in activated sludge flocs [7]. The substrate concentration in the Monod model is the substrate concentration in the bulk, which is not the same as in the floc, since concentration gradients exist inside activated sludge flocs [8]. This is important to understand as a prerequisite for understanding and accepting its analogous use in zero-dimensional biofilm modelling, where the Monod model is used for modelling the macrokinetic behaviour of biofilm systems rather than the intrinsic kinetics in the biofilm [7, 9].

This has implications for the interpretation of the Monod model parameters in activated sludge and biofilm modelling [7]. Essentially it needs to be understood that the values of the Monod model parameters are a function of various processes taking place in and out of the cell aggregates, be it an activated sludge floc or a biofilm. This fact has been nicely illustrated for activated sludge flocs in the literature by the picture of resistances in series [2]: It has been outlined that the value of the half-saturation index is a result of different processes, namely transport in the medium (wastewater), transport in the floc, transport through the membrane, transport in the periplasm, and enzymatic binding and release. All these processes (except enzymatic binding and release) are governed by diffusion. Diffusion can therefore implicitly be taken into account by adapted values of the half-saturation index in the Monod terms of activated sludge models. The same approach has been taken in zero-dimensional biofilm modelling [9].

To recall, the Monod model has been successfully used in state of the art ASMs. In ASMs the Monod model describes the macrokinetic behaviour of activated sludge systems rather than the intrinsic microkinetic behaviour of activated sludge flocs. This means that the substrate concentration in the Monod equation is the substrate concentration in the bulk, not in the activated sludge floc, where concentration gradients exist. This is also true for biofilm systems and the role that biofilm and floc structure has played in this context has been thoroughly discussed in another textbook [7]. Essentially biofilm structure has been strongly emphasized in the biofilm modelling community by proposing one-dimensional (1D), two-dimensional (2D) and three-dimensional (3D) models, whilst state of the art activated sludge models do not take floc structure into account, i.e., they are zero-dimensional (0D).

A UNIFIED VIEW OF GROWTH PROCESSES IN BIOLOGICAL WASTEWATER TREATMENT

Two types of growth processes can be distinguished in biological wastewater treatment, i.e., suspended and attached growth. The latter is also referred to as biofilm growth. In suspended growth systems the biomass grows and moves freely in the reactor whilst in biofilm systems, the biomass grows attached to a surface, i.e., the substratum. From a technical point of view suspended and attached growth systems are different: In suspended growth systems like activated sludge the biomass follows the wastewater flow, whilst in biofilm systems the biomass is retained in the reactor by the substratum. However, from a microbiological point of view both types of biomass can be regarded as cell aggregates, activated sludge flocs being an aggregated form of biomass in suspension and biofilm being an aggregated form of biomass on a surface. Some researchers consider suspended aggregates therefore as mobile biofilms [10]. Technical applications of cell aggregates in wastewater treatment include thus activated sludge (classic, sequencing batch, and membrane reactors), granular sludge reactors, moving bed biofilm reactors, rotating biological contactors and trickling filters. Because all these systems use cell aggregates, they can be approached and described in a unified way from a modelling perspective. This approach can be successful, if the Monod model is correctly applied for both activated sludge and biofilm systems as described above.

WHY THINK AND MODEL BIOFILMS IN LAYERS?

Historically biofilms are often thought in layers forming a thin homogenous film on the substratum. This picture of biofilms is inadequate, since advanced microscopic examination tools, namely the confocal laser scanning microscope (CLSM), have revealed that biofilms have a complex heterogeneous three-dimensional structure with cell clusters, pores, and channels (see [11] and references in there). The complex structure of biofilms has been a driving force towards two- and three-dimensional (2D and 3D) biofilm models. Further it is known that diffusion is an important process in biofilm systems [12] and the application of Fick's laws of diffusion appeared to be a must in biofilm modelling. The application of Fick's laws of diffusion requires spatial dimensionality, a circumstance that has been an additional driving force towards 2D and 3D models. However, 2D and 3D models have not found application in engineering practice because they are too complex. Contrarily a gap has opened in the biofilm modelling community between research and engineering practice over the past decades, whilst moving from 1D to 2D and 3D models [13, 14]. The simplest way of using Fick's laws of diffusion in a model is in 1D, and the author believes that this is why 1D biofilm models have become the most prominent for engineering application, in an attempt to move towards simplified models and narrow the gap between research and engineering practice. In line with the trend towards simplified biofilm models 0D models have recently been proposed and applied to pilot scale plants [9, 15]. Pure simulation studies using 0D approach can also be found in earlier literature [16, 17]. In the 0D modelling approach the Monod model is used to model the macrokinetic behaviour of biofilm systems, analogous to the use of the Monod model in ASMs (see above). The 0D modelling approach overcomes thus a fundamental problem, i.e., model the structure of cell aggregates like biofilms and activated sludge flocs. Further it has been demonstrated that the Monod model substitutes an explicit description of diffusion using Fick's laws in both activated sludge and biofilm modelling, if applied correctly and if the parameters of the Monod model are interpreted properly. There is therefore no need any more to think biofilms in layers just to enable the application of Fick's laws of diffusion in its simplest possible way. Further the development of 2D and 3D biofilm models has in the author's opinion been a misdirection towards biofilm models for engineering practice, although these models may be interesting in biological wastewater treatment research.

A QUALITY ASSESSMENT OF ZERO-DIMENSIONAL MODELS FOR BIOLOGICAL WASTEWATER TREATMENT

The golden rule of modelling says that "models should be as simple as possible and as complex as needed" [18]. This rule holds for both suspended and attached growth modelling and the 0D approach respects this rule. There are further quality criteria which are fulfilled by 0D models:

- 0D models are based on the Monod model and hence on existing knowledge, since the Monod model has found wide application in biochemical engineering. Therefore the Monod model has turned out to be applicable to cell aggregates as found in activated sludge and biofilm systems.
- The practicability of 0D activated sludge models (ASMs) has been shown in engineering practice on full scale plants and the practicability of 0D biofilm models has been demonstrated on pilot-scale plants. The 0D approach is thus practicable.
- 0D models conform the principle of parsimony and resist Ockham's razor in contrast to 1D, 2D, and 3D models.
- The 0D modelling approach can be inspiring. Numerous variations of ASMs have been developed and applied in engineering practice, whilst 0D biofilm models can be adapted and applied to various biofilm systems and validated on full-scale plants.

Zero-dimensional models appear thus to be superior to dimensional models in biological wastewater engineering. - Although a dimensional activated sludge model has been proposed [19], the dimensional activated sludge modelling approach has not found its place in engineering practice.

FUTURE DIRECTIONS

Zero-dimensional (0D) activated sludge models (ASMs) are established tools in engineering practice, since they have found such a wide application. Zero-dimensional models have also been proposed for biofilm systems, however the biofilm modelling community seems to stick to classic one-dimensional models for application in engineering practice, although the application and calibration of 1D models is difficult: Calibration protocols for 1D biofilm models are still under development (see for example [20] and [21]), although 1D biofilm models have already been proposed in the 1970s [22]. The author therefore calls for a shift to 0D biofilm modelling in the biofilm modelling community, since the 0D approach has been proven successful in activated sludge modelling and since both suspended and attached growth processes are based on similar cell aggregates with concentration gradients. Overall the 0D approach appears to be appropriate for modelling biological wastewater treatment systems, since it passes all the quality criteria presented above. The predictive power of zero-dimensional biofilm models should be further evaluated and compared to dimensional models using more data from laboratory reactors and from both pilot- and full-scale plants.

REFERENCES

[1]   Monod, J. (1949). The Growth of Bacterial Cultures. *Annual Review of Microbiology*, 3: 371-394.

[2]   Arnaldos, M., Amerlinck, Y., Rehman, U., Maere, T., Van Hoey, S., Naessens, W. and Nopens, I. (2015). From the affinity constant to the half-saturation index: Understanding conventional modelling concepts in novel wastewater treatment processes. *Water Research*, 70: 458-470.

[3]   Liu, Y., Lin, Y.-M. and Yang, S.-F. (2003). A Thermodynamic Interpretation of the Monod Equation, *Current Microbiology*, 46: 233-234.

[4]   Liu, Y. (2007) Overview of some theoretical approaches for derivation of the Monod equation, *Applied Microbiology and Biotechnology*, 73: 1241-1250.

[5]   Ashby, M. T. (2007) Appreciating Formal Similarities in the Kinetics of Homogeneous, Heterogeneous, and Enzyme Catalysis, *Journal of Chemical Education*, 84 (9): 1515-1519.

[6]   Henze, M., Gujer, W., Mino, T. and van Loosdrecht, M. C. M. (2000) IWA Task Group on Mathematical Modelling for Design and Operation of Biological Wastewater Treatment, Activated Sludge Models ASM1, ASM2, ASM2d and ASM3, *Scientific and Technical Report No. 9*, IWA Publishing, London, England.

[7]   Plattes, M. (2009) The Role of Biofilm and Floc Structure in Biological Wastewater Treatment Modelling. Chapter 10 in *Biochemical Engineering*. Ed.: Fabian E. Dumont and Jack A. Sacco, Nova Science Publishers, New York: 245-255.

[8]   Li, B. and Bishop, P. L. (2004) Micro-profiles of activated sludge floc determined using microelectrodes, *Water Re*search, 38: 1248-1258.

[9]   Plattes, M., Henry, E. and Schosseler, P. M. (2008) A zero-dimensional biofilm model for dynamic simulation of moving bed bioreactor systems: Model concepts, Peterson matrix, and application to a pilot-scale plant. *Biochemical Engineering Journal*, 40 (2): 392-398.

[10] Flemming, H.-C., Wingender, J., Szewzyk, U., Steinberg, P., Rice, S, A. and Kjelleberg, S. (2016) Biofilms: an emergent form of bacterial life. *Nature Reviews Microbiology*, 14 (9): 563-575.

[11] Wanner, O. (1995) New Experimental Findings and Biofilm Modelling Concepts, *Water Science & Technology*, 32 (8): 133-140.

[12] Harremoës, P. and Henze, M. (1997) Biofilters, In: *Wastewater Treatment – Biological and Chemical Processes*, 2nd ed., Henze, M., Harremoës, P., la Cour Jansen, J. and Arvin, E. (ed.), Springer, Berlin: 143.

[13] Noguera, D. R., Okabe, S. and Picioreanu, C. (1999), Biofilm Modeling: Present Status and Future Directions, *Water Science & Technology*, 39 (7): 273-278.

[14] Morgenroth, E., Van Loosdrecht, M. C. M. and Wanner, O. (2000) Biofilm models for the practitioner, *Water Science & Technology*, 41 (4-5): 509-512.

[15] Volcke, E. I. P., Sanchez, O., Steyer, J.-P., Dabert, P. and Bernet, N. (2008) Microbial population dynamics in nitrifying reactors: Experimental evidence explained by a simple model including interspecies competition. *Process Biochemistry*, 43: 1398-1406.

[16] Chen, L.-M. and Chai, L.-H. (2005) Mathematical model and mechanisms for biofilm wastewater treatment systems. *World Journal of Microbiology & Biotechnology*, 21: 1455-1460.

[17] Buriticá, K. L., Trujillo, S. C., Acosta, C. D. and Diaz, A. G. (2015) Dynamical Analysis of a Continuous Stirred-Tank Reactor with the Formation of Biofilms for Wastewater Treatment. *Mathematical Problems in Engineering*, Article Number 512404: 10 pages.

[18] Wuertz, S. and Falkentoft, C. M. (2003) Modeling and Simulation: Introduction, in: *Biofilms in Wastewater Treatment - An Interdisciplinary Approach*, Wuertz, S., Bishop, P. L. and Wilderer, P. A. (ed.), IWA Publishing, London: 3.

[19] Martins, A. M. P., Picioreanu, C., Heijnen, J. J. and Van Loosdrecht, M. C. M. (2004) Three-Dimensional Dual-Morphotype Species Modeling of Activated Sludge Flocs, *Environmental Science and Technology*, 38 (21): 5632-5641.

[20] Barry, U., Choubert, J.M., Canler, J.P., Hédui,t H., Robin, L. and Lessard P. (2012) A calibration protocol of a one-dimensional moving bed bioreactor (MBBR) dynamic model for nitrogen removal. *Water Science and Technology*, 65 (7), 1172-1178.

[21] Rittmann, B. E., Boltz, J. P., Brockmann, D., Daigger, G. T., Morgenroth, E., Sørensen, K. H., Takács, I., van Loosdrecht, M. and Vanrolleghem, P. A. (2018) A framework for good biofilm reactor modelling practice (GBRMP). *Water Science and Technology*, in press.

[22] Wuertz, S. and Falkentoft, C. M. (2003) Modelling and simulation: Introduction. In: *Biofilms in Wastewater Treatment – An Interdisciplinary Approach.* Wuertz, S., Bishop, P. L. and Wilderer, P. A. (ed.), IWA Publishing, London, UK: 6.

In: The Activated Sludge Process
Editor: Benjamin Lefèbvre

ISBN: 978-1-53615-202-9
© 2019 Nova Science Publishers, Inc.

Chapter 4

THE EFFECT OF COMBINED MICROWAVE AND HYDROGEN PEROXIDE PRETREATMENT ON SLUDGE CHARACTERISTICS AND OXIDATION STATUS OF WASTE ACTIVATED SLUDGE

Herald Wilson Ambrose[1,2], Ligy Philip[3],
Tushar Kanti Sen[1,*] and G. K. Suraishkumar[2]
[1]Chemical Engineering, WASM: Minerals, Energy & Chemical
Engineering,, Curtin University, Western Australia
[2]Department of Biotechnology, Indian Institute of Technology (IIT)-
Madras, Chennai, India
[3]Department of Civil Engineering, Indian Institute of Technology (IIT)-
Madras, Chennai, India

* Corresponding Author Email: t.sen@curtin.edu.au.

ABSTRACT

Anaerobic digestion of waste activated sludge helps in sludge mass reduction, biogas production and sludge stabilization. This process involves a series of biological reactions viz; hydrolysis, acedogenesis, acetogenesis and methanogenesis that are carried out by respective microbial communities. The limiting factor for anaerobic digestion is slow hydrolysis rate. Various sludge pre-treatment methods such as advanced oxidation process (AOP), thermal, microwave, electrochemical, biological, ultra-sonication and hydrodynamic cavitation have been used to improve slow hydrolysis, dewaterability and reduce overall sludge volume. Pre-treatments have variable effects on sludge characteristics such as pH, chemical oxygen demand, total organic carbon, total protein and carbohydrate, zeta potential, suspended and volatile solid concentration. The pre-requisite of anaerobic digestion is to enhance solubility of particulate organic matter and facilitate digestion by hydrolytic bacterium. Conventional heat transfer in thermal treatment solubilizes insoluble organics and helps in odour control. Dielectric polarization caused by microwaves is responsible for solubilizing complex organic molecules, thereby effective in breaking EPS (Extracellular polymeric substances) and release of intracellular substances.

Hydrogen peroxide (H_2O_2) is a strong oxidant, used in advanced oxidation process of waste water treatment. H_2O_2 undergoes Fenton type oxidation, leading to generation of hydroxyl radicals [OH^-] and superoxide [O_2^-] that oxidise various target compounds and causes cell lysis & EPS dissolution. Decomposition rate of H_2O_2 to [OH^-] is dependent on temperature. To accelerate [OH] generation, H_2O_2 is treated along with O_3, UV, ultrasound, thermal and microwave processes. In the current research work, different microwave power outputs (100W to 1100W) and time (1-3 min) were optimised for sludge solubilisation without evaporation loss in waste activated sludge from two different sources (NWAS & ATEAS). The variable effects of pre- treatments on EPS fraction, cellular oxidative stress and solubilisation of both sludges were evaluated to understand the impact of sludge complexity in pre-treatment. The optimum condition for maximum sludge solubilisation is 450W & 1% H_2O_2/TS and 880W & 1% H_2O_2/TS for NWAS and ATEAS, respectively. 30-40% higher sludge solubilisation (SCOD/ TCOD) was observed in combined microwave and H_2O_2 oxidation treatment compared to individual treatment in NWAS. The combined treatment produced 8 and 4.1-fold higher s[OH^-] and s[O_2^-] respectively compared to the control experiment. It has been found that higher oxidative stress has corresponded to significant reduction in volatile suspended solids and chemical oxygen demand. Significant reduction in zeta potential in combined pre-treatment indicates an enhanced disaggregation and flocculation tendency in waste activated sludge. Thereby combined microwave and oxidation treatment improves dewaterability of sludge compared to individual treatments.

Keywords: microwave oxidant pre- treatment, waste activated sludge, reactive oxygen species, extracellular polymeric substances (EPS) and dewaterability

INTRODUCTION

Activated sludge process involves the use of aerobic treatment of wastewater, which yields significant production of sludge with high water content and volatile solid fraction. This results in huge volumes of sludge solids ultimately affecting transportation and disposal costs. Waste activated sludge also threatens environment due to presence of heavy metals, organic pollutants, pathogens and odors, that cause secondary environmental pollution (S. Yu et al. 2013). With increasing regulations in industrial nations for sludge disposal, there is a need for residual solid reduction and improved dewaterability (Pérez-Elvira, Nieto Diez, and Fdz-Polanco 2006). Over the decades, various physical, chemical and biological treatment methods have been studied for sludge minimization. Anaerobic digestion is a process widely applied as a sludge minimization technique, for its robustness and lower operational costs. It achieves significant reduction in sludge volume and pathogen levels. It also produces biogas, which could be utilized as an electricity source for treatment plants. Various research groups are constantly studying to further improve the efficiency of anaerobic digestion.

The decomposition of organic matter and biogas production are caused by microbiological process whereby anaerobic microorganisms metabolize complex organic matter and release methane, carbon dioxide, hydrogen sulfide, nitrogen and other trace elements (Shen et al. 2015). The process involves 4 sequential biological reactions namely hydrolysis, acidogenesis, acetogenesis and methanogenesis carried out by distinct microbial community (Ennouri et al. 2016). Hydrolysis is considered the major rate limiting step, due to the slow rate of mass transfer from particulate organic matter from solid phase to liquid phase by hydrolytic bacterium (Tyagi and Lo 2013). Due to the slow hydrolysis of particulate matter, methanogenic process is also limited and directly inflicting on biogas production(Q. Zhang, Hu, and Lee 2016). Waste activated sludge is profiled with complex sludge flocs that are composed of extracellular polymeric substances (EPS) and microbial cells. These EPS fractions are majorly composed of proteins, carbohydrates, lipids and nucleic acids excreted from microbial metabolism. These complexes form three dimensional matrix structures inhibit anaerobic degradability of sludge by affecting mass transfer, microbial aggregate stability and decreases the rate of hydrolysis(Sheng, Yu, and Li 2010). It also has a significant effect on dewaterability of sludge, because of its hydrophilic nature. To enhance hydrolysis rate of anaerobic digestion, destruction of sludge flocs is fundamental and various pretreatment methods have been studied to achieve this (Carrère et al. 2010).

To enhance sludge hydrolysis and achieve residual solid reduction, various pretreatments (Carrère et al. 2010) have been proposed in literature including oxidation process, thermal, microwave, acid and alkaline, surfactants, electrochemical, biological and enzymatic, ultrasonication and hydrodynamic cavitation. Thermal pretreatment breaks down sludge flocs and causes microbial cell lysis. Conventional heating at temperature 160°C to 180°C have been reported to have increased sludge solubilisation and biogas production (Carrère et al. 2010; Ennouri et al. 2016). Conventional heating is less economical as it takes longer time when compared to microwave heating (Yeneneh 2014). This may be attributed to sludge age and floc complexity as heat diffuses from one molecule to another in conventional heating. Microwave treatment has a unique heating mechanism, in which the dielectric polarization caused by MWs is responsible for heating water solutions (X. Zhang et al. 2006). As sewage comprises of water, organic substances, proteins and microbial biomass, microwave heating causes rapid movement of water dipole which eventually causes heating and microbial cell rupture (Jing Zhang et al.

2016). This unique heating mechanism also affects protein and organic substance solubility and helps in sludge dewaterability (Liu, Yu, et al. 2016; Yeneneh et al. 2015). Microwave irradiation causes solubilisation of complex biomolecules that comprise sludge flocs. Mariuz et al., 2013 reported 32% sludge solubilisation with microwave irradiation at a sludge temperature about 90°C (Kuglarz, Karakashev, and Angelidaki 2013). The research also reported a linear correlation of protein solubilization and sludge temperature under microwave irradiation at ≥70°C. Microwave irradiation has an athermal effect, that causes greater damage to microbial cells compared to conventional heating (Eskicioglu et al. 2007). Although the athermal death kinetics of microbial cells or the mechanism involved in microwave irradiation is not clearly established. Eskicioglu et al. reported 20% higher biogas yield in microwave irradiation compared to conventional heating (Eskicioglu et al. 2018). Various researches on 'microwave- pretreatment' have reported increased organic removal, sludge solubilisation and biogas production (Park et al. 2004; Kuglarz, Karakashev, and Angelidaki 2013; Appels et al. 2013). Microwave irradiation increased SCOD by 1.8 to 4.0 fold, soluble protein by 1.1- 1.8 fold, soluble carbohydrate by 3.2- 14.1 fold, in pure cultures of gram positive and gram negative bacterium studied by Zhou et al. (B. W. Zhou et al. 2010). The study also states that microwave irradiation is effective in disintegrating gram- negative organisms which is in high population in municipal sewage sludge. Optimization of microwave power output, treatment time and power intensity is paramount to account for cost effectiveness compared to other methods. Partial evaporation of volatile solids have been observed while microwaving sludge at near boiling temperatures (Kuglarz, Karakashev, and Angelidaki 2013).

Another environment friendly pretreatment method that can increase solubilisation is advanced oxidation process (AOP). AOP causes microbial rupture and release cellular debris into soluble phase. It leads to break down of sludge flocs and EPS fractions thereby helpful in accelerating sludge hydrolysis. AOP involves generation of reactive oxygen species (ROS) such as hydroxyl radicals [OH^-] and superoxides [O_2^-] which cause DNA damage (Slupphaug, Kavli, and Krokan 2003), cell membrane permeabilization (Alvarez et al. 1978) and induction of stress enzymes (Mishra, Noronha, and Suraishkumar 2005). Disruption of EPS fractions by ROS is responsible for improved dewaterability and sludge settleability (Zhen et al. 2014). Ozonation and Fenton reaction by hydrogen peroxide addition are the most established AOP treatments. A classic example of AOP is Fenton reaction, wherein highly reactive [OH·] species are generated from a redox reaction between ferrous ion and hydrogen peroxide (equation 1). [OH·] radical concentration is also dependant on ferrous, ferric and superoxide concentrations in biological systems (equation 2). Fe (II) is regenerated from Fe (III) by a chain mechanism which produces more [OH·] radical as shown by the equation.

$$Fe^{2+} + H_2O_2 \rightarrow Fe^{3+} + HO\cdot + HO^- \tag{1}$$

$$Fe^{3+} + O_2^{\cdot-} \rightarrow Fe^{2+} + O_2 \tag{2}$$

[O_2^-] radicals on the other hand, can be formed from one electron reduction mechanism from molecular oxygen (Prousek 2007) as shown by equation 3.

$$O_2 \rightarrow O_2^{\cdot-} \rightarrow H_2O_2 \rightarrow HO\cdot \rightarrow H_2O \tag{3}$$

Fenton reaction is a preferable oxidation system in sludge pretreatment, because of the abundance of ferrous ion in sewage sludge. AOP is advantageous for its rapid reaction but a high dose is detrimental as it oxidises solubilised nutrients and affects dewaterability (Erden and Filibeli 2010; Yeom et al. 2002). Determining the concentrations of intracellular superoxide and hydroxyl radicals helps in better understanding of AOP mechanism. Various researchers have reported effect of Fenton, photo- Fenton reactions on anaerobic digestion. Most recently Heng et al. have reported 49% increase of COD removal achieved in anaerobic digestion of photo- Fenton pretreated system (Heng et al. 2017). The effect of Fenton based oxidation treatment on dewaterability has shown to reduce CST to about 98.25% in waste activated sludge (Zhen et al. 2014).

Ozonation has been shown to have significant positive effect on sludge disintegration, biochemical methane potential and sludge dewaterability (Erden and Filibeli 2010). AOP in general has been studied extensively for its effect in anaerobic digestion and dewaterability. The mechanism of reactive species generation and their effect in sludge solubilisation is not clearly established.

Recent attraction in sludge pre-treatment is the use of hybrid techniques. Various researchers have combined two pre-treatment techniques for improving digestion performance, cost effectiveness, treatment time and eco- friendliness of pretreatment. Ozonation combined with ultrasonication (Weavers and Hoffmann 2000), microwave combined with ultrasonication (Yeneneh et al. 2015), thermo-chemical (Penaud, Delgenès, and Moletta 1999), microwave combined with oxidation (Eskicioglu et al. 2008) are some of the combined methods. The application of two pre-treatments together has improved sludge disintegration, biogas production, biopolymer release and sludge dewaterability (Gogate and Pandit 2004). Combination of two treatment techniques not only improves the performance individually, but also takes part in overcoming drawbacks in individual treatments. The major driving force in advanced oxidation process is the release of free radicals in high concentrations. The dissociation of hydrogen peroxide into free radicals is affected by temperature (Eskicioglu et al. 2008). Hydrogen peroxide can be 'activated' to produce more hydroxyl radicals by ozone, UV light, transition metals (Fe^{2+}) or microwave treatment. Degradation of synthetic humic substances in drinking water by combined ultrasonication and hydrogen peroxide has been investigated by Chemat et al. (Chemat et al. 2001). The study reports the importance of hydrogen peroxide dosage, as it could act both as an initiator and scavenger. Various researches that involve combined treatment of ultrasonication and hydrogen peroxide have been done for degradation of pollutants and enhancement of anaerobic digestion (Gogate and Pandit 2004). In all studies, the decomposition rate of hydrogen peroxide and active concentration of hydroxyl radicals during reaction are key components. Hydroxyl radicals oxidize compounds with a high reaction rate of order $109M-1$ s-1. Combined pre- treatment of microwave and Fenton- type oxidation has been studied recently by various researchers for sludge solubilisation, biogas production and dewaterability (Neyens et al. 2003; Eswari et al. 2016; Eskicioglu et al. 2008; Liu et al. 2017).

Eskicioglu et al., studied the synergistic effect of microwave combined with hydrogen peroxide for sludge solubilisation, anaerobic digestion performance and dewaterability in thickened waste activated sludge (Eskicioglu et al. 2008). The study mentioned that combined pre- treatment caused effective solubilisation, but not-so-good anaerobic digestion performance compared with microwave treated and control samples. One of the most important factor for a better anaerobic digestion performance is 'optimum pre- treatment' in terms of solubilisation, negligible refractory compound formation and extracellular polymeric substances breakdown.

In the current research work, the effect of microwave & hydrogen peroxide on various sludge parameters have been studied in individual and combination on two different sludges. Our major objective is a better understanding of treatment effects on sludge solubilisation, oxidative stress, biopolymer solubilisation & dewaterability. Studying two different sludges from various sources gives more insight on sludge behaviour towards various treatments. Our study also includes specific reactive oxygen species analysis, that would enhance the interpretation of pre-treatment effectives and mechanism of action.

MATERIALS AND METHODS

Waste Activated Sludge Samples

Waste activated sludge samples were collected from Nesapakkam waste water treatment plant (NWWTP), Chennai, India. NWWTP has a treatment capacity of 54 MLD, and treats one fifth of city's total municipal waste. NWWTP has a primary treatment unit (screen chamber and primary clarifier) followed by activated sludge plant. NWAS (Nesapakkam waste activated sludge) was collected from secondary effluent of aeration tank.

After collection, NWAS was characterized immediately and stored at 4°C until further use (Table 1). The effect of different microwave power output and combined treatment in sludge solubilisation was studied at Environmental and Water Resources Engineering, IIT Madras. Specific reactive oxygen species production was measured at Department of Biotechnology, IIT Madras. ATEAS (Australia thickened excess activated sludge) was collected from a waste water treatment plant located in Western Australia. ATEAS was characterised immediately after collection and stored at 4°C until further use (Table 1). The effect of microwave power intensity, power density and combined pre-treatment efficiency in sludge solubilisation and EPS fraction was studied at Curtin University.

Table 1. Initial characterization of NWAS and ATEAS

S. No	Parameter	NWAS	ATEAS
1	pH	7.37	6.8
3	TCOD (mg/l)	8448 ± 135	39000 ± 150
4	SCOD (mg/l)	785 ± 90	4000 ± 190
5	TS (mg/l)	8630 ± 75	27200 ± 220
6	VS (mg/l))	7909 ± 95	21900 ± 180
7	TSS (mg/l)	8350 ± 70	26065 ± 165
8	VSS (mg/l)	7600 ± 290	20050 ± 75

Analytical Methods

The sludge samples were characterised for pH, total and soluble chemical oxygen demand (TCOD, SCOD), total solids (TS), volatile solids (VS), total suspended solids (TSS), volatile suspended solids (VSS), total organic carbon (TOC), soluble biomolecules (carbohydrate and protein), specific intracellular reactive oxygen species and zeta potential. pH, temperature and conductivity were measured using WP-81 and WP-90 conductivity/TDS/temperature meter equipped with a glass electrode. Soluble fraction of sample was acquired by centrifugation of sludge samples at 5000 rpm for 15 minutes, followed by filtration using 45μm filter paper. Total and soluble chemical oxygen demands were measured using oxidation by HACH COD reagent and analysed in ORION UV/VIS spectrometer. TS, TSS, VS, and VSS are measured using APHA standard methods. TOC was measured using a TOC analyser. Soluble protein was measured using BCA protein assay. Soluble carbohydrate was measured using phenol- sulfuric acid method for 96- well microtiter plate (Masuko et al. 2005). Zeta potential was measured using a zeta potential analyzer.

Reactive Oxygen Species Quantification

Specific hydroxyl radical concentration was measured using aminophenyl fluorescein dye. The dye is oxidized by [OH⋅] radical to form a fluorescent product which is measured in a fluorescent spectrometer at 490 nm excitation wavelength and 515 nm emission wavelength. 1ml of sludge was centrifuged at 5000 rpm for 10 minutes at 4°C and supernatant was removed. The pellet was washed and resuspended in 1ml phosphate buffer (pH 7.4). APF was added to the suspension to a final concentration of 10μM and incubated for 30 minutes at room temperature. The fluorescence is measured in a Perkin Elmer fluorescent spectrometer. Concentration vs fluorescence standard plot was made using varying concentrations of ironsulfate heptahydrate and hydrogen peroxide. Specific [O_2^-] radical concentration was measured using dihydroethidium dye. Its fluorescent product hydroxyl ethidium was measured in a fluorescent spectrometer at 470 nm excitation wavelength and 570 nm emission wavelength. 1ml of sludge was centrifuged at 5000 rpm for 10 minutes at 4°C and supernatant was removed. The pellet was washed and resuspended in 1ml phosphate buffer (pH 7.4). DHE was added to a final concentration of 10μM. A standard plot for superoxide concentration was generated using Xanthine/Xanthine Oxidase system. Specific reactive oxygen species is determined using equation 4.

$$Specific\ ROS\ (\mu moles/\text{TS}) = \frac{[ROS]\mu M}{Total\ solids\ (\frac{mg}{l})} \tag{4}$$

Microwave Treatment

The microwave equipment used is a household microwave oven provisioned with adjustable power and time settings. Operating frequency of microwave oven is 2450 MHz and power output range (100W to 1100W). 30 ml of NWAS and ATEAS samples were taken in beakers of constant dimension and microwave irradiated for variable time (30s, 60s, 120s and 180s). The beakers were loosely capped to avoid loss of sample by evaporation. Sample volume was measured before and after treatment.

Oxidant Treatment

Waste activated sludge samples were treated with hydrogen peroxide to cause Fenton type oxidation reaction. Samples were treated with two concentrations of hydrogen peroxide viz, 0.5% H_2O_2/TS and 1% H_2O_2/TS prepared from a 30% (V/V) H_2O_2 purchased from Sigma Aldrich. H_2O_2 dosage is greatly affected by catalase activity of sludge. In order to deactivate catalase, sludge samples were preheated first to a temperature of 80°C for 5 minutes, then followed by hydrogen peroxide addition (Y. Wang, Wei, and Liu 2009).

EPS Fraction Extraction

Extracellular polymeric substances (EPS) fractions were isolated with a modified protocol of Huang et al. (Huang et al. 2016). Briefly, 10 ml of sample was taken and centrifuged at 5000 rpm for 5 minutes. The supernatant liquid collected is the 'Slime' fraction. The pellet was completely resuspended in 10 ml of Milli-Q water, by vortex mixer and centrifuged at 5000 rpm for 10 minutes. The supernatant collected is the EPS fraction. Both slime and EPS fractions are stored at -20°C until further analysis.

RESULTS AND DISCUSSION

Effect of Microwave Power Intensity, Density and Time on Sludge Solubilisation

The effects of different microwave power output on NWAS solubilisation in shown in Figure 1. Microwave irradiation causes rapid realignment of water dipoles towards oscillating electromagnetic wave and produces energy in the form of heat. Exposure of sludge to extreme microwave irradiation is detrimental as it vaporises water and thickens sludge. Vaporization decreases both TCOD and SCOD, thereby the percentage of complex biomolecule dissolution into soluble phase is lessened at 800W compared to 600W and 450W treatment as seen in the Figure 1b. It also causes vaporization of organic molecules and releases toxic fumes into environment. For 2 minutes of microwave treatment of NWAS, 450W showed least evaporation with 22.5% sludge solubilisation (SCOD/TCOD) (Figure 1b).

Previous studies have reported that breakdown of microbial cells is majorly caused by thermal effects caused by microwave exposure (Vela and State 1979). Although certain studies have reported athermal cell lysis, the mechanism of athermal cell lysis is not very much established (Eskicioglu et al. 2007). Various studies have shown enhancement in anaerobic digestion, dewaterability and pathogen destruction by means of microwave irradiation of waste activated sludge (Eskicioglu et al. 2007; Yeneneh et al. 2015; Park et al. 2004; N. Wang and Wang 2016). Some research works have also reported that microwave irradiation does not have significant effect in biosolids solubilisation and anaerobic digestion performance compared to conventional heating operated at same temperature (80-160°C) (Mehdizadeh et al. 2013).

In the current research work, we observed (20-24) % COD solubilisation with sole microwave treatment for both 450W and 600W power outputs. We couldn't observe a significant increase in COD solubilisation in NWAS treated with 100W microwave, compared to control. The increase in COD solubilisation from 100W to 450W microwave output is almost linear and concurrent with previous research data (Yeneneh et al. 2017). Yeneneh et al., have shown 18% of COD solubilisation at 640W microwave treatment for 3 minutes (Yeneneh et al. 2017). Eskicioglu et al. reported linear SCOD/TCOD ratios upon microwave heating at 50, 75 and 95°C (Eskicioglu et al. 2007). Microwave power of 700W for 3 minutes yielded 21% COD solubilisation in a palm oil mill effluent, which has an initial COD concentration of 50000 mg/l and also 4000 mg/l of oil & grease (Saifuddin and Fazlili 2009). Yu et al., have reported that 1.5 to 2% sludge disintegration was achieved by 900W microwave irradiation for 60s in a secondary sludge. Further treatment time has caused decrease in dewaterability (Q. Yu et al. 2009). The optimum microwave treatment for maximum COD solubilisation differs depending on various sludge characteristics. Comparing previous studies, we can infer that sludge with higher COD load, takes more microwave power output or time to break down complex sludge flocs that could be achieved with lesser microwave power in a lesser COD load.

The effect of microwave irradiation in sludge solubilisation on VSS/TSS is given in Figure 1a. A minimum of 74% VSS/TSS is achieved from 90% VSS/TSS in raw NWAS, in 600W microwave power. Haug et al., reported 55.2% VSS/TSS from 64.2% raw waste activated sludge upon thermal pre-treatment at 175°C (Haug et al. 1978). The effect of microwave irradiation time on suspended solid solubilisation have been studied previously (I. Byun et al. 2014; I. G. Byun et al. 2018). Higher time of exposure leads to boiling and evaporation. In the present work, we studied the effect of various microwave power outputs in sludge solubilisation for a constant time. The solubilisation of NWAS was observed to increase linearly from 100W to 600W. At 800W, no significant solubilisation was observed in both TSS and VSS. Observing the trend in SCOD and VSS/TSS, 800W microwave irradiation didn't improve the solubilisation. Exposure of NWAS to a high microwave output (800W) causes boiling effect and causes evaporation of water and some volatile organics. This could be the reason for saturation in suspended solid solubilisation at 800W. The optimum microwave power is one in which maximum particulate organics solubilizes into soluble phase without any evaporation by boiling. Yet another disadvantage of treating sludge at boiling temperature is odour emission.

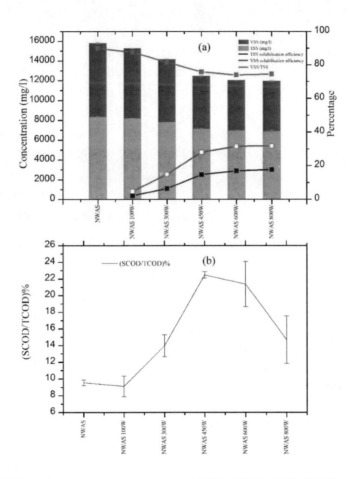

Figure 1. Effect of microwave power outputs (100W, 300W, 450W, 600W and 800W) in NWAS solubilisation. (a) TSS & VSS solubilisation profile of NWAS (b) COD solubilisation profile of NWAS.

Effect of Combined Pre-Treatment in Sludge Solubilisation

To enhance sludge solubilisation and biogas production, microwave treatment has been studied in combination with acid, alkali, ultrasonication and H_2O_2 (Liu, Wei, et al. 2016; Yeneneh et al. 2015; Doğan and Sanin

2009; Eskicioglu et al. 2008). Previous researchers were confined to study the synergistic effect of combined treatment in overall anaerobic digestion performance and biogas production. In the current research, we evaluated effect of combined microwave and H_2O_2 treatment on sludge characteristics in both NWAS and ATEAS. NWAS solubilisation by combined microwave (450W for 120s) and H_2O_2 (0.5% and 1% per TS) treatment was studied. The dosage of hydrogen peroxide in a combined pre- treatment is important, in order to achieve maximum solubilisation of particulate organic substances without causing production of refractory compounds that would mitigate anaerobic digestion. Liu et al. achieved 35-45% sludge solubilisation with a microwave power output of 600W (100°C) and with varying hydrogen peroxide dosage (0.2g, 0.6g and 1g H_2O_2/g TS) (Liu et al. 2017). This study reported the inhibitory effect of residual hydrogen peroxide in overall anaerobic digestion and biogas production. Previously Eskicioglu et al., reported that combined microwave and H_2O_2 (1g H_2O_2/g TS) dosage resulted in 25% SCOD/TCOD ratio, 19% reduction in TS but had lower first order mesophilic biodegradation and methane yield compared to sole microwave treatment and control waste activated sludge (Eskicioglu et al. 2008). The study states, this could be because of slowly biodegradable refractory compounds that are generated during combined treatment. Wang et al., studied the effect of H_2O_2 dosing strategy in a combined microwave-oxidant treatment with H_2O_2 doses ranging from 0.1-4 H_2O_2/TCOD (Y. Wang, Wei, and Liu 2009). Recently Inan et al. reported 2.5% (w/v) H_2O_2 combined with 300W (2.5 minute) microwave treatment in degradation of lignocellulosic biomass (Inan, Turkay, and Akkiris 2016). Eswari et al observed 50% TS solubilisation in a combined pre-treatment (MW + 0.1 to 1mg H_2O_2/g SS) in dairy waste activated sludge (Eswari et al. 2016). Zhang et al. used a dosage of 0.2g H_2O_2/g TS in combination with 600W microwave and studied anaerobic digestion efficiency (Junya Zhang et al. 2017). Neyens et al., studied peroxidation effect in sludge dewatering by using a dosage of 25g H_2O_2/kg DS (Neyens et al. 2003). Lo et al. reported 39% VS reduction and 34-45% TCOD reduction with 1.2% H_2O_2/% TS (Lo et al. 2017).

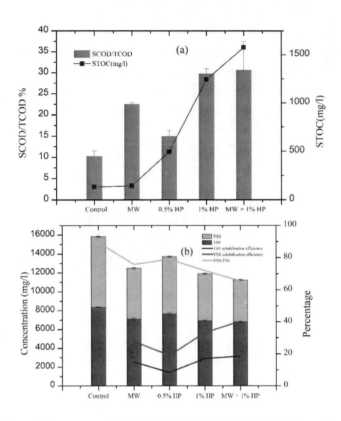

Figure 2. Effect of microwave (450W/120s), 0.5% & 1% H₂O₂/TS & combined (450W/120s & MW+ 1% H₂O₂/TS) in NWAS solubilisation. (a) COD solubilisation profile of NWAS (b) TSS & VSS solubilisation profile of NWAS. HP- Hydrogen peroxide.

Most of the researchers used total solids as a standard for optimising H_2O_2 dosing. Some of them have also reported negative effects of residual H_2O_2 and reduction in anaerobic digestion performance (Eskicioglu et al. 2008; Liu et al. 2017). This is possibly due to high dosage. During our study, the dosage of 1g H_2O_2/g TS completely bleached NWAS, even with slow addition and mixing. We observed the same result in ATEAS as well. In our study we used 0.5% & 1% (w/w) hydrogen peroxide per total solids. By combined treatment (MW+ 1% H_2O_2/TS), 30-40% higher SCOD solubilisation was observed compared to 22% solubilisation in sole microwave treatment (Figure 2a). We observed almost equal amount of solubilisation in sole 1% H_2O_2/TS treatment. Higher AOP efficiency is dependent on inactivation of catalase enzyme, before subjecting it to oxidative stress (Liu et al. 2017). Catalase catalyses the conversion of hydrogen peroxide to water and oxygen (equation 5). Catalase plays a major role in protecting cells from formation of reactive oxygen species under oxidative stress. In the current study NWAS and ATEAS were exposed to 80°C for 5 minutes for catalase inactivation.

$$2H_2O_2 \rightarrow 2H_2O + O_2 \tag{5}$$

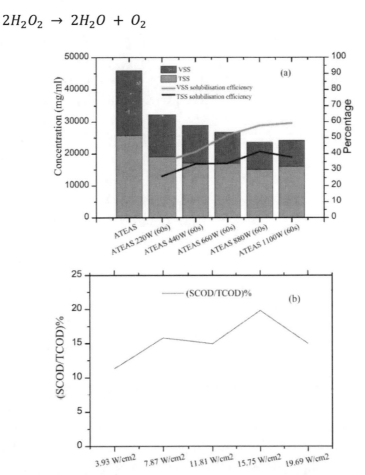

Figure 3. Effect of combined (MW + H₂O₂) with variable power intensities in ATEAS solubilisation. (a) TSS & VSS solubilisation profile of ATEAS for constant contact time (b) COD solubilisation profile of ATEAS.

The highest VSS solubilisation efficiency (40%) was observed in combined (MW+ 1% H_2O_2/TS) treatment. 0.5% H_2O_2/TS & 1% H_2O_2/TS treatment yielded 19% and 33% VSS solubilisation. The synergistic effect of combined pre- treatment has caused generation of ROS species such as hydroxyl and superoxide radicals, therefore increasing solubilisation. The effect of combined microwave (different power intensities) and 1% H_2O_2/TS treatment on ATEAS solubilisation is given in Figure 3. ATEAS has higher COD load compared to NWAS and we observe significant difference in optimum treatment condition. The highest COD solubilisation (22%) is observed at 880W microwave power output (15.75 W/cm^2) (Figure 3b). We observed 15% COD solubilisation in both 440W and 1100W microwave treatment. A decrease in solubilisation at higher microwave intensities has been observed in both NWAS and ATEAS. The possible reason for this could be 'boiling' effect caused in microwave treatment and subsequent evaporation of organics. The VSS solubilisation followed a similar trend to SCOD solubilisation (Figure 3a). Around 60% VSS solubilisation was observed in 1100W+1% H_2O_2/TS and yielded the least VSS/TSS % (51%). Comparing the sludge solubilisation profile of NWAS and ATEAS, we observe ATEAS requires high microwave output (880W) at constant 1% H_2O_2/TS concentration, to achieve maximum solubilisation. This indicates that sludge with high solid content and COD load (Table 1) require more microwave power to pass through complex sludge flocs and cause microbial cell breakdown.

Effect of Pre-Treatment in Biopolymer Solubilisation

Extra-cellular polymeric substances (EPS) of activated sludge are composed of complex mixture of biopolymers such as carbohydrates, proteins, lipids, humic substances and nucleic acids. EPS are excreted by aerobic and anaerobic microorganisms, produced from cellular metabolism, cellular lysis and hydrolysis of macromolecules. EPS forms a complex 3-D matrix around sludge flocs and affect its physico-chemical properties, substrate/product mass transfer, surface charge and flocculation properties of microbial aggregates. Carbohydrates and proteins are major components of EPS. In our current research, we analysed the release of carbohydrate and proteins into soluble phase, upon microwave irradiation and oxidative treatment in ATEAS (Figure 4). ATEAS was treated with varying microwave intensities along with 1% H_2O_2/TS addition. Soluble carbohydrate and soluble protein fraction are extracted from EPS. As seen in Figure 4, the amount of protein release is 3 times higher than amount of carbohydrate released. The highest amount of protein release was observed in 7.88 W/cm^2 (440W/60s) combined with 1% H_2O_2/TS. As the treatment intensity is increased beyond this point, there is a slight decrease in the soluble protein concentration. A same trend is observed in carbohydrate release. 6.6 mg/ml carbohydrate release was achieved in the combined treatment 7.88 W/cm^2 (440W/60s) with 1% H_2O_2/TS, and further power intensity has decreased the carbohydrate release. EPS forms a colloidal interphase between microbial aggregates and bulk of liquid. The production of EPS particularly proteins, is elevated in presence of toxic heavy metals and higher BOD load, as organisms produce more EPS for their protection against harsh environment. The production of higher fraction of protein in ATEAS, goes in accordance with Sheng et al's report on EPS behaviour in sludge (Sheng, Yu, and Li 2010) and solubilisation of biopolymers by microwave and conventional heating by Mehdizadeh eta l (Mehdizadeh et al. 2013). The decrease in protein and carbohydrate concentration at higher microwave intensity could be due to Maillard reaction and caramelization (Eskicioglu et al. 2007). At temperatures above 80°C, reducing sugars and amino acids can polymerize, form melanoidins and transform to particulate phase. Maillard reaction occurs when cooking food rich in protein and carbohydrate at higher temperature with sudden dehydration. With higher COD and BOD load, the complexity of waste activated sludge increases. This causes EPS fractions to be rich in proteins and carbohydrates as seen in ATEAS. And Microwave irradiation is a perfect environment for maillard reaction to happen, than a conventional heating. In our experiments, we observed evaporation of water at higher microwave intensities in ATEAS, leading to thickened and caramelised giant flocs. Some researchers that have worked with microwave treatment also have reported similar decrease in biopolymer concentration at higher microwave intensities (Eskicioglu et al. 2007; Toreci, Kennedy, and Droste 2010). Formation of maillard products can subsequently affect anaerobic digestion efficiency. Dwyer et al., has reported that thermal hydrolysis process operated at 165°C, had poor biodegradability in anaerobic digestion because of formation of melanoidins (Dwyer et al. 2008). In spite of high solubilisation by pre-treatment the overall anaerobic digestion performance can decline because of maillard reactions. So optimisation of pre- treatment techniques to avoid melanoidin formation is much needed. At microwave intensities 11.82 W/cm^2 (600W & 1% H_2O_2/TS) & 15.76 W/cm^2 (880W & 1% H_2O_2/TS), we observed equivalent amount of biopolymer solubilisation (Figure 4a). At 1100W (& 1% H_2O_2/TS), the solubilisation of biopolymer has significantly decreased. A concurrent result is observed in COD solubilisation at this pre- treatment. With that data, we conclude 1100W (& 1% H_2O_2/TS) is not suitable for achieving maximum ATEAS solubilisation efficiency without having negative impact on anaerobic digestion. The optimum combined treatment condition for ATEAS is (880W & 1% H_2O_2/TS), where maximum COD solubilisation and significant EPS destruction is achieved. Another important parameter to be considered in microwave irradiation is contact time. Some researchers have worked on this aspect (Koupaie and Eskicioglu 2016; Al-jabiry and Zouari 2012). Koupaie et al., have reported that slow heating ramp rates (3°C/min) to a temperature of 160°C solubilized significantly higher amount of biopolymers compared to fast heating ramp rates (6°C/min) & (11°C/min) (Koupaie and Eskicioglu 2016). Eskicioglu et al., reported 1.3°C/min ramp rate to have better solubilisation capacity compared to 1.4°C/min & 1.2°C/min (Eskicioglu et al. 2007). In our study, we observed that faster heating rates is not advantageous, which can be observed from decreased solubilisation at 1100W (&1% H_2O_2/TS) treatment (Figure 4b). As seen from figure, the release of protein increased with increase in treatment time and attained maximum at 440W for 2 minutes. A maximum of 6.6 mg/ml soluble carbohydrate is released in (440W/60s) with 1% H_2O_2/TS treatment. There is a non-linearity in release of proteins and carbohydrates into soluble phase through pre-treatment. Treatment at 180 s caused a sharp decline in both biopolymer solubilisation. With a combined treatment of 660W & 1% H_2O_2/TS, maximum dissolution of protein was observed in 60s, further treatment time decreased dissolution and also caused vaporization of sample.

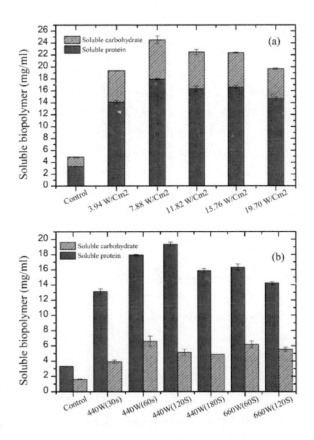

Figure 4. Effect of combined (MW + 1%H_2O_2/TS) with variable power intensities in EPS solubilisation. (a) Biopolymer solubilisation at various power intensities (b) Biopolymer solubilisation at various microwave contact time of ATEAS.

Effect of Pre-Treatment on Oxidative Stress of Sludge

In advanced oxidation process (AOP), the generation of [OH·] and [O_2^-] radicals can degrade sludge flocs, disrupt EPS and thus improve sludge settleability and dewaterability. Hydrogen peroxide is a strong oxidant, and exposure to microwave accelerates its dissociation into hydroxyl radicals. The highly reactive hydroxyl radicals can oxidise target compounds with reaction rates of the order of 10^9 M^{-1} s^{-1}. A significant role is played by another radical- superoxide; it replenishes the concentration of hydroxyl radials through hydrogen peroxide formation (Equation 3). The specific levels of intracellular reactive species, hydroxyl and superoxide, are direct indicators of oxidative stress (Menon, Balan, and Suraishkumar 2013). Therefore, to understand the oxidative stress effects in sludge microbes, intracellular quantification of both hydroxyl & superoxide radical is required. The role of ROS in oxidative pretreatment of activated sludge has been investigated by some researchers (Gong et al. 2015; A. Zhang, Wang, and Li 2015). The generation of hydroxyl radicals in sludge because of microwave treatment has been studied earlier, but the quantification is limited. Quantification of ROS species by fluorescent probes is an attractive technology in case of biology and medicine. Fluorescent probes are highly specific to target reactive species and can even help real- time monitoring in living cells (Zhuang et al. 2014). Radical scavengers THIO and DABCO were used to study effect of [OH·] and [$O_2^{·-}$] radicals in removal of endocrine-disrupting compounds in waste activated sludge by calcium peroxide treatment (A. Zhang, Wang, and Li 2015). Zhang et al., reported that [OH·] and [$O_2^{·-}$] radicals are the functional ROS species involved in EDC removal. Also, [OH·] radical was reported to be the most potent oxidiser (A. Zhang, Wang, and Li 2015). Gong et al., quantified hydroxyl radical concentration by terephthalic acid (TA) trapping in an ultrasonic assisted Fenton oxidation of excess sludge (Gong et al. 2015). Sarkar et al., quantified intracellular superoxide radicals using hydroxyethidium fluorescence in *Bacillus subtilis* under pH and temperature stress (Sarkar and Suraishkumar 2011). Jose et al., quantified intracellular hydroxyl radical concentration using aminophenyl fluorescein (APF) in a gut microbe, *Enterococcus durans* (Jose, Bhalla, and Suraishkumar 2018).

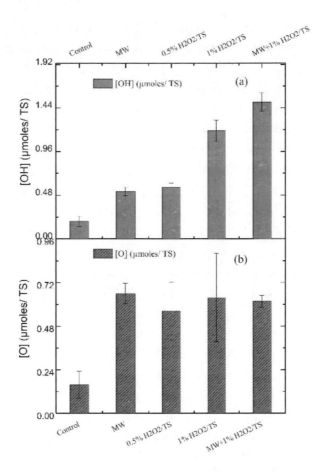

Figure 5. Effect of microwave (450W/120s), 0.5% & 1% H_2O_2/TS & combined (450W/120s & MW+ 1% H_2O_2/TS) in sROS generation in NWAS (a) s[OH·] for control NWAS, microwave, H_2O_2 & combined (MW + H_2O_2) treatment (b) s[O_2^-] for control NWAS, microwave, H_2O_2 & combined (MW + H_2O_2) treatment.

In our current research we quantified intracellular hydroxyl and superoxide radicals using aminophenyl fluorescein (APF) and dihydroethidium (DHE), respectively according to established reports (Sarkar and Suraishkumar 2011; Jose, Bhalla, and Suraishkumar 2018). The fast- reacting hydroxyl radicals have a half-life of about 1ns, and direct detection is highly impossible (Buxton et al. 1988). However, the pseudo-steady state levels of reactive species can give significant insights, and are measureable (Menon, Balan, and Suraishkumar 2013). The quantification of fluorescent product (APF*), helps us to quantify [OH·] radicals. The effect of microwave (450W/120s) and oxidant treatment (0.5% and 1% H_2O_2 per TS) on sROS (specific reactive oxygen species) in NWAS was studied (Figure 5). The specific ROS measurement determines the absolute oxidative stress on microbial cells in an activated sludge upon treatment with oxidant and microwave.

Highest s[OH·] concentration (1.5 µmole/g cells) was observed in combined treatment, which was 1.3 fold higher than 1% H_2O_2/TS and 3 fold higher than 0.5% H_2O_2/TS (Figure 5a). The microwave treatment accelerates the decomposition of H_2O_2 to hydroxyl radicals. Microwave treatment and 0.5% H_2O_2/TS treatment, almost have equivalent amount of s[OH·]. When we compare s[O_2^-] radical measurement, we observe the highest concentration of superoxides (0.66 µmole/g cells) in pure microwave treatment (Figure 5b). This could be a reason for higher flux in s[OH·] concentration in microwave treatment, as superoxide can eventually form hydroxyl radical (equation 3). Shaoying et al., reported elevated superoxide scavenging enzyme activity in apple juice treated at 720W and 900W from 75-100s (S. Zhang and Zhang 2014). The study also reported that microwave induction can enhance 'chelating' and 'reducing' property of apple juice. And also the study justifies maillard reaction as a result of increase in reducing capability of apple juice. In our study we also observe decreasing biopolymer release at elevated microwave treatment (880W & 1100W), that could be justified by maillard reaction (Figure 4b) (Eskicioglu et al. 2007). Studying superoxides, expands our knowledge in better understanding the oxidative stress caused by microwave and oxidant treatment of waste activated sludge. Sludge solubilisation in NWAS goes concurrent with the oxidative stress caused by hydroxyl and superoxide radicals. The combined treatment, H_2O_2 (0.5%), H_2O_2 (1%) and MW have produced 8, 3, 6.2 & 2.7 fold higher s[OH·] respectively compared to control. The combined treatment, H_2O_2 (0.5%), H_2O_2 (1%) and MW have produced 4.1, 3.7, 4.2 & 4.4 fold higher s[O_2^-] respectively compared to control.

Effect of Pre-Treatment in Settleability

Surface forces at interface between sludge and suspending liquid become an important factor for understanding flocculation phenomenon. Microbial suspension in sludge forms a colloidal phase interacting with surrounding liquid which is majorly water. Activated sludge flocs carry negative surface charge in a range between -10 mV to -20mV, due to ionization of anionic functional groups (X. S. Jia 1996). Zeta potential (ZP) is the potential difference between surface of solids (immersed in a liquid) and the bulk of the liquid. ZP is measured by the particle velocity induced when a potential difference is applied along a capillary tube containing the sample. ZP is helpful in understanding sedimentation and flocculation phenomenon, and is largely dependent on electrical surface charge on microorganisms. Dependant on cell wall structure, microbial cells have higher or lower surface charge which determine their flocculation properties. ZP of microbial cells vary according to species, for example Escherichia coli -24.96 mV, *Alcaligenes faecalis* -34.24 mV and *Leucobacter* sp -12.78 mV (Xie, Gu, and Lu 2010). To understand how the treatment techniques affect the surface charge of different fractions of activated sludge, NWAS was partitioned into 3 fractions, viz NWAS (solid fraction), slime fraction and LB- EPS (Loosely bound EPS). The zeta potential of untreated NWAS (solid fraction) is -52.37 ± 4.21 mV. This may be due to heavy BOD load in treatment plants, or the process technology or the constituents of waste water (Bennoit 2014). Absorption of charged ions by sludge flocs due to its large surface area can be another reason for the high negative ZP (Tang and Zhang 2014). Higher the zeta potential, poorer is the flocculating tendency of sludge. Increased negative surface charge leads to greater repulsive electrostatic force according to DLVO theory (Xie, Gu, and Lu 2010).

Pre- treatments have yielded -44.87 mV, -37.9 mV, -30.06 mV and -22.41 mV for microwave, 0.5% H2O2/TS, 1% H2O2/TS and combined (MW+1% H2O2/TS) for NWAS (Figure 6). As seen from the Figure 6, the combined pre-treatment has increased the zeta potential from ~ -52 mV to -22 mV. Treatment imparts positive charge on microbial surface which was highly negative (~ -52 mV) in raw NWAS. This could be because of release of divalent cations (Ca^{2+} and Mg^{2+}), due to floc breakage (Chu et al. 2001). These divalent cations are major bridging elements in floc structure. Microwave and oxidative treatment causes major floc deterioration which results in release of these divalent cations. Huang et al., has reported neutralization of negative cell surface charge by Mg^{2+} ions from $Mg(OH)_2$ pre-treatment (Huang et al. 2016). The pre- treatment techniques break down chemical bonds and dissolve complex organic matrix, thereby destabilizing the suspended particles and bring down its surface charge. Higher the oxidative stress, higher is the disaggregation and flocculating tendency of particulate matter in solid fraction of sludge.

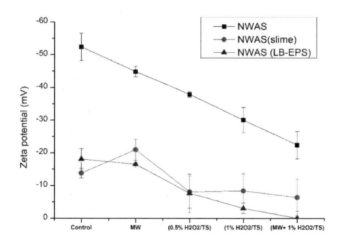

Figure 6. Effect of microwave (400W/120s), 0.5% H_2O_2/TS, 1% H_2O_2/TS & combined (450W/120s & MW+ 1% H_2O_2/TS) in Zeta Potential of various fractions of NWAS.

The extracted LB- EPS and slime fraction have -13mV and -18mV ZP respectively. These fractions are majorly composed of proteins and carbohydrates. Zhang et al., has reported a detailed analysis on various EPS proteins present in aerobic, anaerobic and anoxic sludges and their characteristics by shotgun proteomics (P. Zhang et al. 2015). The isoelectric points and molecular mass analysis of the report indicates that most of the EPS proteins carry a negative charge. Approximately 73% of these proteins were identified as binding and catalytic proteins. The EPS architecture varies from sludge to sludge due to various reasons, such as sludge age, protein diversity, microbial population, pollutant level and aerobic/anaerobic condition. According to our results, the pre- treatments significantly push the ZP towards the positive charge. This explains how LB-EPS and slime enhances the dewaterability of NWAS overall. This result is concurrent with Zhou et al. (X. Zhou 2015). From this result, we conclude that all three fractions respond positively towards dewaterability by addition of various pre- treatments, and combined pre- treatment being best out of all three.

CONCLUSION

In the current research work we analysed effects of microwave and hydrogen peroxide treatment, in individual and combination on various parameters like sludge solubilisation, oxidative status and EPS behaviour in two separate sludges (NWAS & ATEAS). From the research we found that response of sludge to a particular pre- treatment can vary according its microbial density and architecture. An optimised hybrid treatment condition for a particular sludge may not apply for another sludge. The optimised treatment that caused maximum solubilisation in NWAS is 450W and 1% H_2O_2/TS, whereas for ATEAS it is 880W & 1% H_2O_2/TS. And in both treatment conditions, we found combined (MW & H_2O_2) has a better performance compared to individual treatments, in terms of solubilisation efficiency and oxidative stress. Apart from well-established role of hydroxyl radicals in causing oxidation, superoxide radicals also play a crucial role in oxidation process. An effective treatment is one that causes maximum solubilisation without causing evaporation of organics. 'Boiling' temperatures can be detrimental as they lead to formation of refractory compounds that will cause difficulty in hydrolysis step of anaerobic digestion. Pre- treatments impact the surface charge of various fractions of sludge differently. The fraction containing microbial aggregates responded linear to our treatment regime, whereas EPS fraction and slime responded differently due to the presence of proteins and carbohydrates. This work widens the scope of pre- treatment optimisation which can be helpful in achieving maximum anaerobic digestion performance.

ACKNOWLEDGMENTS

The authors would like to thank Chemical Engineering of Curtin University-Perth, Australia and Department of Biotechnology of Indian Institute of Technology, Madras, India for necessary research support and scholarship support under Curtin-IITM collaborative Ph.D. program.

REFERENCES

Al-jabiry, Hayat, and Nabil Zouari. 2012. "Characterization and Anaerobic Digestion of Waste Waters of Poultry Meat Industry and Local Slaughterhouse." *Qatar Foundation Annual Research Forum Proceedings* 4 (3). https://doi.org/10.5339/qfarf.2012.EEPS14.

Alvarez, Juan G, Touchstone Joseph C, Blasco Luis, and Bayard T Storey. 1978. "Spontaneous Hydrogen Superoxide Enzyme Lipid Peroxide Dismutase Peroxidation and Production in Human of Spermatozoa and Superoxide as Major Against Oxygen Toxicity." *Journal of Andrology*, 338–48. https://doi.org/10.1002/j.1939-4640.1987.tb00973.x.

Appels, Lise, Sofie Houtmeyers, Jan Degreve, Jan Van Impe, and Raf Dewil. 2013. "Influence of Microwave Pre-Treatment on Sludge Solubilization and Pilot Scale Semi-Continuous Anaerobic Digestion." *Bioresource Technology* 128: 598–603. https://doi.org/10.1016/j.biortech.2012.11.007.

Bennoit, H. 2014. *Improvement of Separation Processes in Waste Water Treatment by Controlling the Sludge Properties* (January 2001): 1–9. https://doi.org/10.13140/2.1.1086.3685.

Buxton, George V., Clive L. Greenstock, W. Phillips Helman, and Alberta B. Ross. 1988. "Critical Review of Rate Constants for Reactions of Hydrated Electrons, Hydrogen Atoms and Hydroxyl Radicals (OH/O) in Aqueous Solution." *Journal of Physical and Chemical Reference Data* 17 (2): 513–886. https://doi.org/10.1063/1.555805.

Byun, I G, J H Lee, J M Lee, J S Lim, and T J Park. 2018. *Evaluation of Non-Thermal Effects by Microwave Irradiation in Hydrolysis of Waste-Activated Sludge*, no. October: 742–49. https://doi.org/10.2166/wst.2014.295.

Byun, Imgyu, Jaeho Lee, Jisung Lim, Jeongmin Lee, and Taejoo Park. 2014. *Impact of Irradiation Time on the Hydrolysis of Waste Activated Sludge by the Dielectric Heating of Microwave*. 19 (1): 83–89.

Carrère, H., C. Dumas, A. Battimelli, D. J. Batstone, J. P. Delgenès, J. P. Steyer, and I. Ferrer. 2010. "Pretreatment Methods to Improve Sludge Anaerobic Degradability: A Review." *Journal of Hazardous Materials* 183 (1–3): 1–15. https://doi.org/10.1016/j.jhazmat.2010.06.129.

Chemat, F, P G M Teunissen, S Chemat, and P V Bartels. 2001. "Sono-Oxidation Treatment of Humic Substances in Drinking Water." *Ultrasonics Sonochemistry* 8: 247–50.

Chu, C. P., Bea Ven Chang, G. S. Liao, D. S. Jean, and D. J. Lee. 2001. "Observations on Changes in Ultrasonically Treated Waste-Activated Sludge." *Water Research* 35 (4): 1038–46. https://doi.org/10.1016/S0043-1354(00)00338-9.

Doğan, Ilgin, and F. Dilek Sanin. 2009. "Alkaline Solubilization and Microwave Irradiation as a Combined Sludge Disintegration and Minimization Method." *Water Research* 43 (8): 2139–48. https://doi.org/10.1016/j.watres.2009.02.023.

Dwyer, Jason, Daniel Starrenburg, Stephan Tait, Keith Barr, Damien J Batstone, and Paul Lant. 2008. "Decreasing Activated Sludge Thermal Hydrolysis Temperature Reduces Product Colour, without Decreasing Degradability." *Water Research* 42: 4699–4709. https://doi.org/10.1016/j.watres.2008.08.019.

Ennouri, Hajer, Baligh Miladi, Soraya Zahedi Diaz, Luis Alberto Fernandez Guelfo, Rosario Solera, Moktar Hamdi, and Hassib

Bouallagui. 2016. "Effect of Thermal Pretreatment on the Biogas Production and Microbial Communities Balance during Anaerobic Digestion of Urban and Industrial Waste Activated Sludge." *Bioresource Technology* 214: 184–91. https://doi.org/10.1016/j.biortech.2016.04.076.

Erden, G, and A Filibeli. 2010. "Ozone Oxidation of Biological Sludge: Effects of Disintegration, Anaerobic Biodegradability and Filterability." *Environmental Progress & Sustainable Energy* 30 (3): 377–383. https://doi.org/10.1002/ep.10494

Eskicioglu, Cigdem, Ronald L Droste, Kevin J Kennedy, Cigdem Eskicioglu, Ronald L Droste, and Kevin J Kennedy. 2018. *Performance of Anaerobic Waste Activated Sludge Digesters After Microwave Pretreatment.* 79 (11): 2265–73. https://doi.org/10.2175/106143007X176004.

Eskicioglu, Cigdem, Audrey Prorot, Juan Marin, Ronald L. Droste, and Kevin J. Kennedy. 2008. "Synergetic Pretreatment of Sewage Sludge by Microwave Irradiation in Presence of H2O2 for Enhanced Anaerobic Digestion." *Water Research* 42 (18): 4674–82. https://doi.org/10.1016/j.watres.2008.08.010.

Eskicioglu, Cigdem, Nicolas Terzian, Kevin J. Kennedy, Ronald L. Droste, and Mohamed Hamoda. 2007. "Athermal Microwave Effects for Enhancing Digestibility of Waste Activated Sludge." *Water Research* 41 (11): 2457–66. https://doi.org/10.1016/j.watres.2007.03.008.

Eswari, Parvathy, S. Kavitha, S. Kaliappan, Ick Tae Yeom, and J. Rajesh Banu. 2016. "Enhancement of Sludge Anaerobic Biodegradability by Combined Microwave-H2O2pretreatment in Acidic Conditions." *Environmental Science and Pollution Research* 23 (13): 13467–79. https://doi.org/10.1007/s11356-016-6543-2.

Gogate, Parag R, and Aniruddha B Pandit. 2004. "A Review of Imperative Technologies for Wastewater Treatment II : Hybrid Methods." *Advances in Environmental Research* 8: 553–97. https://doi.org/10.1016/S1093-0191(03)00031-5.

Gong, Changxiu, Jianguo Jiang, De'an Li, and Sicong Tian. 2015. "Ultrasonic Application to Boost Hydroxyl Radical Formation during

Fenton Oxidation and Release Organic Matter from Sludge." *Scientific Reports* 5 (11419). https://doi.org/10.1038/srep11419.

Haug, Roger T, David C Stuckey, James M Gossett, and Perry L Mccarty. 1978. "Effect of Thermal Pretreatment on Digestibility and Dewaterability of Organic Sludges." *Water Pollution Control Federation* 50 (1): 73–85. https://doi.org/10.2307/25039508.

Heng, Gan Chin, Mohamed Hasnain Isa, Jun Wei Lim, Yeek Chia Ho, and Ali Akbar Lorestani Zinatizadeh. 2017. "Enhancement of Anaerobic Digestibility of Waste Activated Sludge Using Photo-Fenton Pretreatment." *Environmental Science and Pollution Research*, 1–12. https://doi.org/10.1007/s11356-017-0287-5.

Huang, Cheng, Jia Lai, Xiuyun Sun, Jiansheng Li, Jinyou Shen, Weiqing Han, and Lianjun Wang. 2016. "Enhancing Anaerobic Digestion of Waste Activated Sludge by the Combined Use of NaOH and Mg(OH)2: Performance Evaluation and Mechanism Study." *Bioresource Technology* 220: 601–8. https://doi.org/10.1016/j.bio rtech.2016.08.043.

Inan, H, O Turkay, and C Akkiris. 2016. "Microwave and Microwave-Chemical Pretreatment Application for Agricultural Waste." *Desalination and Water Treatment* 57 (6): 2590–96. https://doi.org/ 10.1080/19443994.2015.1069217.

Jose, Steffi, Prerna Bhalla, and G K Suraishkumar. 2018. "Oxidative Stress Decreases the Redox Ratio and Folate Content in the Gut Microbe, Enterococcus Durans (MTCC 3031)." *Scientific Reports* 8: 1–7. https://doi.org/10.1038/s41598-018-30691-4.

Koupaie, E Hosseini, and C Eskicioglu. 2016. "Conventional Heating vs Microwave Sludge Pretreatment Comparison under Identical Heating/ Cooling Profiles for Thermophilic Advanced Anaerobic Digestion." *Waste Management* 53: 182–95. https://doi.org/10.1016/j.wasman. 2016.04.014.

Kuglarz, Mariusz, Dimitar Karakashev, and Irini Angelidaki. 2013. "Microwave and Thermal Pretreatment as Methods for Increasing the Biogas Potential of Secondary Sludge from Municipal Wastewater

Treatment Plants." *Bioresource Technology* 134: 290–97. https://doi. org/10.1016/j.biortech.2013.02.001.

Liu, Jibao, Ruilai Jia, Yawei Wang, Yuansong Wei, Junya Zhang, Rui Wang, and Xing Cai. 2017. "Does Residual H2O2 Result in Inhibitory Effect on Enhanced Anaerobic Digestion of Sludge Pretreated by Microwave-H2O2 Pretreatment Process?" *Environmental Science and Pollution Research* 24 (10): 9016–25. https://doi.org/10.1007/s11356-015-5704-z.

Liu, Jibao, Yuansong Wei, Kun Li, Juan Tong, Yawei Wang, and Ruilai Jia. 2016. "Microwave-Acid Pretreatment : A Potential Process for Enhancing Sludge Dewaterability." *Water Research* 90: 225–34. http://dx.doi.org/10.1016/j.watres.2015.12.012.

Liu, Jibao, Dawei Yu, Jian Zhang, Min Yang, Yawei Wang, Yuansong Wei, and Juan Tong. 2016. "Rheological Properties of Sewage Sludge during Enhanced Anaerobic Digestion with Microwave-H2O2 Pretreatment." *Water Research* 98: 98–108. https://doi.org/10.1016/j.watres.2016.03.073.

Lo, Kwang Victor, Ruihuan Ning, Cristina Kei, and Yamamoto De Oliveira. 2017. *Application of Microwave Oxidation Process for Sewage Sludge Treatment in a Continuous-Flow System.* 143 (9): 1–9. https://doi.org/10.1061/(ASCE)EE.1943-7870.0001247.

Masuko, Tatsuya, Akio Minami, Norimasa Iwasaki, Tokifumi Majima, Shin Ichiro Nishimura, and Yuan C. Lee. 2005. "Carbohydrate Analysis by a Phenol-Sulfuric Acid Method in Microplate Format." *Analytical Biochemistry* 339 (1): 69–72. https://doi.org/10.1016/j.ab.2004.12.001.

Mehdizadeh, Seyedeh Neda, Cigdem Eskicioglu, Jake Bobowski, and Thomas Johnson. 2013. "Conductive Heating and Microwave Hydrolysis under Identical Heating Profiles for Advanced Anaerobic Digestion of Municipal Sludge." *Water Research* 47: 5040–51. http://dx.doi.org/10.1016/j.watres.2013.05.055.

Menon, Kavya R, Ranjini Balan, and G K Suraishkumar. 2013. "Stress Induced Lipid Production in Chlorella Vulgaris : Relationship With

Specific Intracellular Reactive Species Levels." *Biotechnology and Bioengineering* 110 (6): 1627–36. https://doi.org/10.1002/bit.24835.

Mishra, Surabhi, S. B. Noronha, and G. K. Suraishkumar. 2005. "Increase in Enzyme Productivity by Induced Oxidative Stress in Bacillus Subtilis Cultures and Analysis of Its Mechanism Using Microarray Data." *Process Biochemistry* 40 (5): 1863–70. https://doi.org/10.1016/j.procbio.2004.06.055.

Neyens, E., J. Baeyens, M. Weemaes, and B. De Heyder. 2003. "Pilot-Scale Peroxidation (H2O2) of Sewage Sludge." *Journal of Hazardous Materials* 98 (1–3): 91–106. https://doi.org/10.1016/S0304-3894(02) 00287-X.

Park, B., John Hwa Ahn, Jaai Kim, and S. Hwang. 2004. "Use of Microwave Pretreatment for Enhanced Anaerobiosis of Secondary Slugde." *Water Science and Technology* 50 (9): 17–23.

Penaud, V., J. P. Delgenès, and R. Moletta. 1999. "Thermo-Chemical Pretreatment of a Microbial Biomass: Influence of Sodium Hydroxide Addition on Solubilization and Anaerobic Biodegradability." *Enzyme and Microbial Technology* 25 (3–5): 258–63. https://doi.org/10.1016/S0141-0229(99)00037-X.

Pérez-Elvira, S. I., P. Nieto Diez, and F. Fdz-Polanco. 2006. "Sludge Minimisation Technologies." *Reviews in Environmental Science and Biotechnology* 5 (4): 375–98. https://doi.org/10.1007/s11157-005-5728-9.

Prousek, Josef. 2007. "Fenton Chemistry in Biology and Medicine." *Pure and Applied Chemistry* 79 (12): 2325–38. https://doi.org/10.1351/pac200779122325.

Saifuddin, N, and S. A. Fazlili. 2009. "Effect of Microwave and Ultrasonic Pretreatments on Biogas Production from Anaerobic Digestion of Palm Oil Mill Effleunt." *American Journal of Engineering and Applied Sciences* 2 (1): 139–46.

Sarkar, Pritish, and G. K. Suraishkumar. 2011. "PH and Temperature Stresses in Bioreactor Cultures: Intracellular Superoxide Levels." *Industrial and Engineering Chemistry Research* 50 (23): 13129–36. https://doi.org/10.1021/ie200081k.

Shen, Yanwen, Jessica L. Linville, Meltem Urgun-Demirtas, Marianne M. Mintz, and Seth W. Snyder. 2015. "An Overview of Biogas Production and Utilization at Full-Scale Wastewater Treatment Plants (WWTPs) in the United States: Challenges and Opportunities towards Energy-Neutral WWTPs." *Renewable and Sustainable Energy Reviews* 50: 346–62. https://doi.org/10.1016/j.rser.2015.04.129.

Sheng, Guo Ping, Han Qing Yu, and Xiao Yan Li. 2010. "Extracellular Polymeric Substances (EPS) of Microbial Aggregates in Biological Wastewater Treatment Systems: A Review." *Biotechnology Advances* 28 (6): 882–94. https://doi.org/10.1016/j.biotechadv.2010.08.001.

Slupphaug, Geir, Bodil Kavli, and Hans E. Krokan. 2003. "The Interacting Pathways for Prevention and Repair of Oxidative DNA Damage." *Mutation Research - Fundamental and Molecular Mechanisms of Mutagenesis* 531 (1–2): 231–51. https://doi.org/10.1016/j.mrfmmm.2003.06.002.

Tang, Bing, and Zi Zhang. 2014. "Essence of Disposing the Excess Sludge and Optimizing the Operation of Wastewater Treatment: Rheological Behavior and Microbial Ecosystem." *Chemosphere* 105 (34): 1–13. https://doi.org/10.1016/j.chemosphere.2013.12.067.

Toreci, Isil, Kevin J. Kennedy, and Ronald L. Droste. 2010. "Effect of High-Temperature Microwave Irradiation on Municipal Thickened Waste Activated Sludge Solubilization." *Heat Transfer Engineering* 31 (9): 766–73. https://doi.org/10.1080/01457630903501039.

Tyagi, Vinay Kumar, and Shang Lien Lo. 2013. "Microwave Irradiation: A Sustainable Way for Sludge Treatment and Resource Recovery." *Renewable and Sustainable Energy Reviews* 18 (71): 288–305. https://doi.org/10.1016/j.rser.2012.10.032.

Vela, G R, and North Texas State. 1979. "Mechanism of Lethal Action of 2,450-MHz Radiation on Microorganisms." *Applied and Environmental Microbiology* 37 (3): 550–53.

Wang, Nannan, and Peng Wang. 2016. "Study and Application Status of Microwave in Organic Wastewater Treatment – A Review." *Chemical Engineering Journal* 283: 193–214. https://doi.org/10.1016/j.cej.2015.07.046.

Wang, Yawei, Yuansong Wei, and Junxin Liu. 2009. "Effect of H2O2 Dosing Strategy on Sludge Pretreatment by Microwave-H2O2 Advanced Oxidation Process." *Journal of Hazardous Materials* 169 (1–3): 680–84. https://doi.org/10.1016/j.jhazmat.2009.04.001.

Weavers, Linda K, and Michael R Hoffmann. 2000. "Kinetics and Mechanism of Pentachlorophenol Degradation by Sonication, Ozonation, and Sonolytic Ozonation." *Environmental Science & Technology* 34 (7): 1280–85. https://doi.org/10.1021/es980795y.

X. S. Jia, Herbert H. P. Fang and H. Furumai. 1996. "Surface Charge and Extracellular Polymer of Sludge in the Anaerobic Degradation Process." *Water Science and Technology* 34: 309–16.

Xie, Bing, Jidong Gu, and Jun Lu. 2010. "Surface Properties of Bacteria from Activated Sludge in Relation to Bioflocculation." *Journal of Environmental Sciences* 22 (12): 1840–45.

Yeneneh, Anteneh Mesfin. 2014. *Study on Performance Enhancement of Anaerobic Digestion of Municipal Sewage Sludge.* Curtin University.

Yeneneh, Anteneh Mesfin, Ahmet Kayaalp, Tushar Kanti Sen, and Ha Ming Ang. 2015. "Effect of Microwave and Combined Microwave-Ultrasonic Pretreatment on Anaerobic Digestion of Mixed Real Sludge." *Journal of Environmental Chemical Engineering* 3 (4): 2514–21. https://doi.org/10.1016/j.jece.2015.09.003.

Yeneneh, Anteneh Mesfin, Tushar Kanti Sen, Ha Ming Ang, and Ahmet Kayaalp. 2017. "Optimisation of Microwave, Ultrasonic and Combined Microwave-Ultrasonic Pretreatment Conditions for Enhanced Anaerobic Digestion." *Water, Air, and Soil Pollution* 228 (1). https://doi.org/10.1007/s11270-016-3197-0.

Yeom, I T, K R Lee, K H Ahn, and S H Lee. 2002. "Effects of Ozone Treatment on the Biodegradability of Sludge from Municipal Wastewater Treatment Plants." *Water Science and Technology* 46 (4–5): 421–25. http://www.ncbi.nlm.nih.gov/pubmed/12361042.

Yu, Qiang, Hengyi Lei, Guangwei Yu, Xin Feng, Zhaoxu Li, and Zhicheng Wu. 2009. "Influence of Microwave Irradiation on Sludge Dewaterability." *Chemical Engineering Journal* 155 (1–2): 88–93. https://doi.org/10.1016/j.cej.2009.07.010.

Yu, Shuyu, Guangming Zhang, Jianzheng Li, Zhiwei Zhao, and Xiaorong Kang. 2013. "Effect of Endogenous Hydrolytic Enzymes Pretreatment on the Anaerobic Digestion of Sludge." *Bioresource Technology* 146: 758–61. https://doi.org/10.1016/j.biortech.2013.07.087.

Zhang, Ai, Jie Wang, and Yongmei Li. 2015. "Performance of Calcium Peroxide for Removal of Endocrine-Disrupting Compounds in Waste Activated Sludge and Promotion of Sludge Solubilization." *Water Research* 71: 125–39. https://doi.org/10.1016/j.watres.2015.01.005.

Zhang, Jing, Kai Yang, Hongyu Wang, Lian Zheng, Fang Ma, Bin Lv, Jing Zhang, et al. 2016. "Impact of Microwave Treatment on Dewaterability of Sludge during Fenton Oxidation." *Desalination and Water Treatment* 57: 14424–32. https://doi.org/10.1080/19443994.2015.1065765.

Zhang, Junya, Jibao Liu, Yawei Wang, Dawei Yu, Qianwen Sui, Rui Wang, Meixue Chen, Juan Tong, and Yuansong Wei. 2017. "Profiles and Drivers of Antibiotic Resistance Genes Distribution in One-Stage and Two-Stage Sludge Anaerobic Digestion Based on Microwave-H2O2pretreatment." *Bioresource Technology* 241: 573–81. https://doi.org/10.1016/j.biortech.2017.05.157.

Zhang, Peng, Yu Shen, Jin-Song Guo, Chun Li, Han Wang, You-Peng Chen, Peng Yan, Ji-Xiang Yang, and Fang Fang. 2015. "Extracellular Protein Analysis of Activated Sludge and Their Functions in Wastewater Treatment Plant by Shotgun Proteomics." *Scientific Reports* 5 (12041). https://doi.org/10.1038/srep12041.

Zhang, Quanguo, Jianjun Hu, and Duu Jong Lee. 2016. "Biogas from Anaerobic Digestion Processes: Research Updates." *Renewable Energy* 98: 108–19. https://doi.org/10.1016/j.renene.2016.02.029.

Zhang, Shaoying, and Rui Zhang. 2014. "Effects of Microwave Pretreatment of Apple Raw Material on the Nutrients and Antioxidant Activities of Apple Juice." *Journal of Food Processing* 2014: 1–10. http://dx.doi.org/10.1155/2014/824050%0AResearch.

Zhang, Xiwang, Yizhong Wang, Guoting Li, and Jiuhui Qu. 2006. "Oxidative Decomposition of Azo Dye C.I. Acid Orange 7 (AO7) under Microwave Electrodeless Lamp Irradiation in the Presence of

H2O2." *Journal of Hazardous Materials* 134 (1–3): 183–89. https://doi.org/10.1016/j.jhazmat.2005.10.046.

Zhen, Guangyin, Xueqin Lu, Baoying Wang, Youcai Zhao, Xiaoli Chai, Dongjie Niu, and Tiantao Zhao. 2014. "Enhanced Dewatering Characteristics of Waste Activated Sludge with Fenton Pretreatment: Effectiveness and Statistical Optimization." *Frontiers of Environmental Science and Engineering* 8 (2): 267–76. https://doi.org/10.1007/s11783-013-0530-3.

Zhou, Bi Wen, Seung Gu Shin, KwangHyun Hwang, Johng Hwa Ahn, and Seokhwan Hwang. 2010. "Effect of Microwave Irradiation on Cellular Disintegration of Gram Positive and Negative Cells." *Applied Microbiology and Biotechnology* 87 (2): 765–70. https://doi.org/10.1007/s00253-010-2574-7.

Zhou, Xu. 2015. *Improving Dewaterability of Waste Activated Sludge through Advanced Oxidization Process.*

Zhuang, Mei, Changqin Ding, Anwei Zhu, and Yang Tian. 2014. "Ratiometric Fluorescence Probe for Monitoring Hydroxyl Radical in Live Cells Based on Gold Nanoclusters." *Analytical Chemistry* 86: 1829–36. https://doi.org/10.1021/ac403810g.

In: The Activated Sludge Process
Editor: Benjamin Lefèbvre

ISBN: 978-1-53615-202-9
© 2019 Nova Science Publishers, Inc.

Chapter 5

THE EFFECT OF TOXIC CARBON SOURCES ON THE REACTION PROCESS OF ACTIVATED SLUDGE

Changyong Wu[1,2], Min Xu[2,3] and Yuexi Zhou[1,2,+]*

[1]State Key Laboratory of Environmental Criteria and Risk Assessment, Chinese Research Academy of Environmental Sciences, Beijing, China
[2]Research Center of Water Pollution Control Technology, Chinese Research Academy of Environment Sciences, Beijing, China
[3]College of Water Science, Beijing Normal University, Beijing, China

ABSTRACT

Activated sludge technology is the most used option for the treatment of wastewater. The toxic carbon source can cause higher residual effluent dissolved organic carbon than easily biodegraded carbon source in the activated sludge process. Based on the variations of the chemical components of activated sludge, mainly intracellular storage materials

* Corresponding Author Email: changyongwu@126.com.
+ Corresponding Author Email: zhouyuexi@263.net.

(X_{STO}), extracellular polymeric substances (EPS), and soluble microbial products (SMP), the performance and mechanism of toxic carbon on the reaction process of activated sludge can be clarified. In addition, the integrated activated sludge model based on carbon flows can be used to understand the mechanism. In the steady state, the toxic carbon can result in higher microbial cells death rate, decay rate coefficient of biomass, the utilization-associated products (UAP) and EPS formation coefficients, than that of easily biodegraded carbon, indicating that more carbon flows into the extracellular components, such as SMP, when degrading toxic organics. In the non-steady state, the yield coefficient for growth and maximum specific growth rate are very low in the acclimatization stage, while the decay rate coefficient of biomass and microbial cells death rate are relatively high.

1. INTRODUCTION

Activated sludge technology has been widely used in the field of wastewater treatment. The activated sludge process is generally considered to have its origins in aeration experiments. Organic pollutants can be degraded by microorganisms, mainly bacteria, during the activated sludge process and then the wastewater is purified.

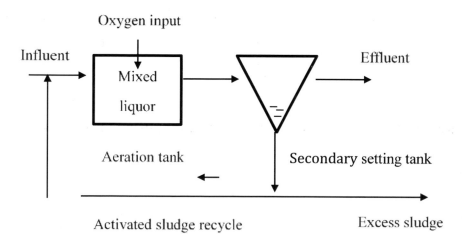

Figure 1. Schematic diagram of the activated sludge process.

2.2. Soluble Microbial Products (SMP)

SMP are generated during the normal metabolic process in the biological wastewater treatment process. Generally, it is recognized that the compositions of SMP are the same as EPS. SMP can be divided into utilization-associated products (UAP) and biomass-associated products (BAP). UAP are associated with substrate metabolism and biomass growth and they are produced at a rate proportional to the rate of substrate utilization, while BAP are associated with biomass decay and are produced at a rate proportional to the concentration of biomass. The release of EPS and cell lysis may be important sources of SMP (Aquino et al., 2006). Therefore, the compositions of the SMP are the same as EPS. In the secondary effluent of biological wastewater treatment systems, it has been reported that the SMP is the majority component of the residual chemical oxygen demand (COD) (Han et al., 2013). In addition, the SMP has a potential effect on water resources contamination (Imai et al., 2002). Therefore, the reduction of SMP is very important to enhance the discharge levels of the final biological effluent, and it attracted many investigations in the field of biological wastewater treatment (Aquino et al., 2004, 2006; Zhang et al., 2010). The factors which affect the formation of SMP have been researched deeply in recent years. Process parameters such as feed strength, hydraulic retention time (HRT), organic load rate (OLR), sludge detention time (SRT), substrate type, biomass concentration, temperature and reactor type can affect SMP production (Barker et al., 1999). It is a more preferable option than increasing the size of the treatment plant, with its associated increase in investment costs and higher operation and maintenance costs.

2.3. Intracellular Storage Products (X_{STO})

X_{STO} is formed during the feast period and it can also supply the carbon source during the famine period of the operation (Laspidou et al., 2002b; Ni et al., 2009). Activated sludge processes can be highly dynamic

In the conventional continuous flow process configuration, as illustrated in Figure 1.1, the activated sludge biomass is separated from the mixed liquid in a secondary settling tank (SST) and the separated sludge is recycled back to the aeration tank. The operating mixed liquid suspended solids (MLSS) concentration may vary in the range of 1500-5000 mg L^{-1}. The active fraction of the MLSS is conveniently approximated for design purposes as its organic component or mixed liquid volatile suspended solids (MLVSS). The MLVSS to MLSS ratio is typically within the range of 0.65-0.85.

2. COMPONENTS OF EPS, SMP AND X_{STO} FOR ACTIVATED SLUDGE

2.1. Extracellular Polymeric Substances (EPS)

In biological wastewater treatment systems, most of the microorganisms are present in the form of microbial aggregates, such as sludge flocs, biofilms, and granules (Sheng et al., 2010). EPS, a complex high-molecular-weight mixture of polymers, are present both outside of cells and in the interior of microbial aggregates. It is recognized that EPS mainly consists of proteins, carbohydrates, polysaccharides, humic substances, etc. (Jørgensen et al., 2017). The content and compositions of the EPS extracted from various microbial aggregates are reported to be heterogeneous (Wingender et al., 1999). The variation in the compositions of the extracted EPS is attributed to many factors, such as culture, growth phase, process parameters, bioreactor types, extraction methods, and analytical tools used (Nielsen et al., 1999). EPS can maintain the structure and skeleton strength of the activated sludge flocs, and it is also the mass transfer pathway during the metabolic processes (Liu et al., 2010; Jørgensen et al., 2017; Shi et al., 2017). Process parameters such as nutrient content, bacterial growth phase, substrate type, and external conditions can affect the EPS production (Sheng et al., 2010).

with respect to the feed regime, especially for sequencing the batch reactor (SBR) (van Loosdrecht et al., 1997) and the enhanced biological phosphorus removal processes (Oehmen et al., 2005). In these processes, the active microorganisms are exposed to significant concentrations of the external substrate only for relatively short periods of time. Internally stored products allow them to take advantage of the dynamic feast-and-famine periods. Microorganisms are capable of quickly storing substrate as internal storage products during feast conditions and then consuming the stored substrate during famine conditions. The process has a strong competitive advantage over microorganisms without such a capacity (Ni et al., 2009).

3. ANALYSES OF EPS, SMP AND X_{STO}

3.1. EPS Measurements

The composition of the EPS matrix in biofilms and activated sludge is reported to be very complex, containing proteins, carbohydrates, nucleic acids, lipids, humic substances etc. Generally, the carbohydrate content is measured using the anthrone method or the phenol–sulfuric acid method. Although the two methods for carbohydrates determination in EPS showed that the two methods yielded similar results, the coefficient of variation for the anthrone method was lower than that for the phenol–sulfuric method (Frolund et al., 1996). The protein content is measured using the Lowry method, the Bradford method, or the total N-content method. The Lowry method has a higher recovery rate than the Bradford method (Frolund et al., 1996). The total N-content method is more accurate, but the procedures are complex. Thus, the Lowry method is frequently applied for protein analyses in EPS characterization. Humic substances are very complex, and there are fewer appropriate methods for measuring their content in EPS. Frolund et al. (1995) proposed a modified Lowry method to determine the humic substance content by correcting the protein interference. The DNA or nucleic acid content is measured using the DAPI fluorescence method

(Frolund et al., 1996), the UV absorbance method (Sheng et al., 2005a), or the diphenylamine method (Liu et al., 2002a). Among these methods, the UV absorbance method is easy to perform, but it is readily interfered with by proteins. The DAPI method for DNA estimation works well, but its procedures are very complex. Therefore, the diphenylamine method could be used more conveniently and widely (Sheng et al., 2010).

The complex compositions make it difficult to analyze the conformation, structure, distribution and functions of EPS. However, progress in analytical chemistry has led to the development of new instruments and techniques for the characterization of EPS. Compared with the conventional scanning electron microscopy (SEM) and transmission electron microscopy (TEM), the innovative methods, such as environmental scanning electron microscopy (ESEM) (Beech et al., 1996), atomic force microscopy (AFM) (van der Aa et al., 2002; Li et al., 2004) and confocal laser scanning microscopy (CLSM) (Zhang et al., 2001) could be used to observe the fully hydrated samples to obtain the original shapes and structures of EPS. The spatial distributions of carbohydrates, proteins and nucleic acids in EPS can also be obtained by CLSM after staining by various fluorescence probes (Staudt et al., 2004). Gas chromatography (GC) and gas chromatography - mass spectrometry (GC-MS) could be used to qualitatively and quantitatively analyze the EPS compositions (Dignac et al., 1998). The spectroscopy, including X-ray photoelectron spectroscopy (XPS) (Dufrene et al., 1996), Fourier transform infrared spectroscopy (FTIR) (Allen et al., 2004; Sheng et al., 2006b), 3-dimensional excitation–emission matrix fluorescence spectroscopy (3D-EEM) (Sheng et al., 2006a), and nuclear magnetic resonance (NMR) (Manca et al., 1996; Lattner et al., 2003) could be used to elucidate the functional groups and element compositions in EPS or microbial aggregates. In addition, due to the high sensitivity, good selectivity, and non-destruction of samples, these spectroscopy techniques could also be used to characterize the adsorption pollutants to EPS from the changes of their functional groups in EPS (Omoike et al., 2004).

3.2. SMP Measurements

Most approaches to analyze SMP fractions involve two steps: extraction and quantification. Several studies have compared the use of different chemical and physical extraction techniques to separate SMP from biomass (Rosenberger et al., 2005; Comte et al., 2006, 2007; Park et al., 2007). Reported extraction techniques include settling (Yu et al., 2009b), centrifugation at different accelerations (Rosenberger et al., 2005), screening (Yu et al., 2009b), and filter paper (Tan et al., 2008). Extraction by sonication after centrifugation has also been evaluated (Ji et al., 2006; Comte et al., 2006; Yu et al., 2009a). SMP extract has generally been quantified immediately after extraction, or stored by freezing (Jin et al., 2004; Henriques et al., 2007). The discrepancies in SMP could result from the use of different extraction techniques.

Little information is available on the influence of the activated sludge matrix on SMP analysis, though there are many natural and anthropogenic chemicals in activated sludge. Matrix effects affecting protein measurement can be seen in data reported by Mehrez et al. (2007), where an unknown dilution ratio of bovine serum albumen (BSA) in filtrate changed its recovery by 17%. Matrix effects were further confirmed by Avella et al. (2010), who compared the classic Lowry method, the corrected Lowry method and a commercial assay kit to analyze BSA in the presence of humic acid in a series of standard solutions. Potvin et al. (2011) compared the analysis of the main SMP components in activated sludge using standard addition (SA) and the traditional standard curve approach (TSC). They found that the TSC approaches commonly used for the quantitation of SMP do not account for matrix effects caused by other chemical materials in activated sludge, can greatly affect both accuracy and recovery of SMP measurements. Analysis of SMP in activated sludge by the SA technique provides a way to compensate for matrix effects, and is thus an improvement over the TSC approaches for proteins, humics, carbohydrates and polysaccharides.

3.3. X$_{STO}$ Measurements

The polyhydroxyalkanoates (PHA) in activated sludge is recognized as X$_{STO}$ (van Loosdrecht et al., 1997; Wu et al., 2016). Normally, the samples were lyophilized, digested, methylated, and extracted with chloroform. Then the samples could be measured by the GC system using a flame ionization detector (FID). Polyhydroxy-butyrate (PHB), polyhydroxy-valerate (PHV), and polyhydroxy-2-methyvalerate (PH2MV) were reported as the standards during the measurement. The PHA (X$_{STO}$) is recognized as the sum of PHB, PHV and PH2MV.

4. SMP PRODUCTION CHARACTERISTICS ON TOXIC CONDITIONS

The toxic substances can affect the metabolism of the activated sludge (Huang et al., 2008). More EPS or SMP than normal will be generated when there is the presence of toxic substances as a response of the microbial cells to environmental stresses (Zhang et al., 2010; Han et al., 2013). Many researchers have investigated the effects of heavy metal ions, such as Cr (VI), Ni (II), Zn (II), and Cu (II) on organic degradation, as well as the SMP production of microorganisms (Aquino et al., 2004; Zhang et al., 2010; Han et al., 2013; Wu et al., 2015; Yan et al., 2015). The results have proved the promotion formation of SMP when there were toxic metals. For example, in the activated sludge system, Han et al. (2013) reported that the SMP content increased slightly when the Zn (II) concentration was below 400 mg L^{-1}. However, the SMP increased significantly when the Zn (II) concentration was between 600 and 800 mg L^{-1}.

Many industrial wastewaters contain toxic organics. Huang et al. (2008) investigated the SMP formation characteristics in SBR systems fed with glucose and phenol. They found the phenol could cause a higher effluent SMP than that of glucose. In order to control the generation of

SMP or EPS, and to improve the effluent quality, understanding the impact of toxic organics on the microbial metabolic processes is very important.

5. VARIATIONS OF EPS, SMP AND X_{STO} WITH TOXIC/NONTOXIC CARBONS

5.1. Experimental Study

Wu et al. (2016) have investigated the effects of carbon sources on the metabolic characteristics when changing from acetate to the toxic carbon source phenol. Acetate is often used as an easily biodegradable carbon source and it can be completely removed by aeration in SBR (Dionisi et al., 2008). The changes and relationships of EPS, SMP, and X_{STO} were examined. In addition, the effects of the toxic organics on the death of the activated sludge cells were also investigated.

Wu et al. (2016) used a sequencing batch reactor (SBR) to investigate the effect of carbon sources on the metabolism of activated sludge. Acetate and phenol, with the COD of 330 to 350 mg L^{-1}, was used as the carbon source, respectively. The COD decreased to 31.2 mg L^{-1} 120 min later, and remained at this level for the next 120 min. According to the COD slope pattern, the first 120 min were the feast period, and the last 120 min were the famine period, or in other words, the endogenous respiration period. Acetate decreased in the initial 120 min with the intracellular storage materials (X_{STO}), extracellular polymeric substances (EPS), and the soluble microbial products (SMP) accumulating to 131.0, 347.5, and 35.5 mg L^{-1}, respectively. In the following 120 min, the internal carbon source was used for the metabolism. Then, X_{STO} and EPS decreased to 124.5 and 340.0 mg L^{-1}, respectively (Figure 2).

When acetate was replaced by phenol, it could not be used at the beginning due to its toxicity (Figure 3). There was almost no decrease of COD during the aeration period. The X_{STO} decreased from 142 mg L^{-1} to 54.6 mg L^{-1} during the aeration period. The EPS had a significant increase, with the highest value of 618.1 mg L^{-1}, which then decreased to 245.6 mg L^{-1} at 240 min.

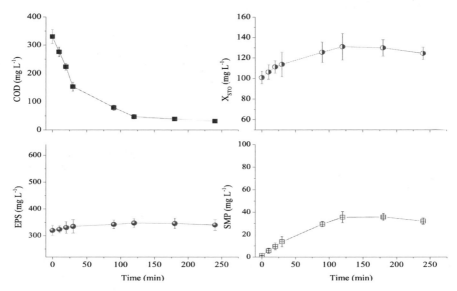

Figure 2. The metabolic characteristics feeding with acetate on stable stage (Wu et al., 2016).

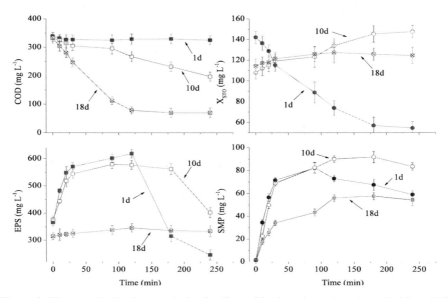

Figure 3. The metabolic characteristics feeding with phenol on phenol day 1, 10 and 18 (Wu et al., 2016).

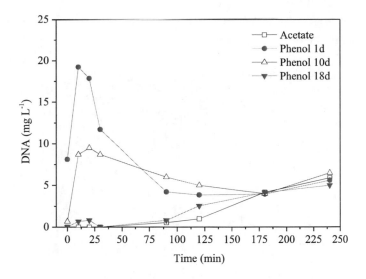

Figure 4. The DNA concentration in the wastewater during different operation periods (Wu et al., 2016).

The rapid increase in the EPS may be caused by the death of the microbial cells and hydrolysis, where some substances, such as proteins and carbohydrates, were released into the bulking liquid (Aquino et al., 2006; Zhang et al., 2010). It has been reported that a portion of the EPS could dissolve into the wastewater, and became SMP (Ni et al., 2009). Compared to the toxic phenol, the EPS and SMP could be used as the external carbon sources for the growth and maintenance of the microbial cells (Huang et al., 2008). The concentration of EPS and SMP decreased in the last 120 min. The phenol was gradually degraded with the acclimation and it can be fully degraded 18 d later. Meanwhile, the usage ratio of the internal carbon source decreased. The effluent SMP feeding phenol was 1.7 times that of feeding acetate. This was due to the fact that the production of SMP could be stimulated in toxic environmental stress conditions, as reported in previous studies (Huang et al., 2008; Yan et al., 2015). According to the mass balance, 3.8% of the carbon could be transferred into X_{STO}, less new cell growth and higher SMP production were conducted compared to acetate as the carbon source. This could be confirmed by the average MLSS on a stable stage during the operation.

The average MLSS was 2477 mg L^{-1} feeding acetate which is higher than that of feeding phenol.

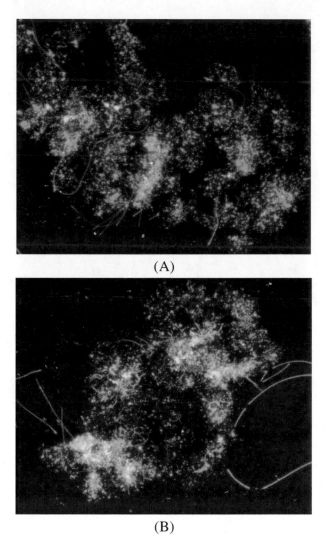

Figure 5. The LIVE/DEAD picture of sludge with acetate on steady stage (A) and phenol on the first day (B) as the carbon sources.

DNA can be released into the wastewater during the endogenous decay period, or under toxic conditions. Therefore, it can be used as an indicator of cell death and lysis (Sheng et al., 2010). The DNA concentration was very low when an external carbon source existed in the first 120 min. However, it increased quickly during the next 120 min, due to the endogenous decay (Figure 4). On phenol day 1, when the carbon source was changed to phenol, the DNA concentration increased sharply during the initial time. This was due to cell death and lysis. The LIVE/DEAD picture can intuitively show the live (green) and dead (yellow) cells of the activated sludge (Figure 5). The calculation also shows that the toxic phenol lead to more cell deaths than acetate (Table 1). The DNA concentration decreased with aeration, as it could be degraded partly by the microorganisms in aeration conditions (Sheng et al., 2010).

Table 1. The proportion of dead cells based on the staining at different stages (samples collected 30 min after aeration)

Carbon source	Acetate	Phenol		
	40 d	1 d	10 d	18 d
Dead cells proportion in activated sludge (%)	17.9	34.5	22.4	18.8

The composition ratios of the EPS and SMP were different, having been fed with different carbon sources in the stable stage. The ratio of protein was the highest of all the components of the EPS, regardless of whether the carbon source was acetic acid or phenol. The percentage of protein and polysaccharide feeding with acetate was slightly higher (5.3%) than that of the feeding with phenol. However, the DNA percentage from feeding phenol was approximately twice that of feeding acetate. DNA comes from the lysis of dead microbial cells (Sheng et al., 2010). The overall concentration of each component in SMP was from high to low as follows: proteins, polysaccharides, humics, and DNA. This was the same as the EPS components' concentration. Due to the toxicity of the phenol for the microbial cells, the SMP concentration was significantly higher than that of the acetate feeding period. Therefore, it could increase the final effluent COD concentration. Finally, the effluent COD feeding phenol was 2.2 times higher than that of feeding acetate.

5.2. Modeling Reaction of the Activated Sludge

To promote the development of mathematical models for biological wastewater treatment systems and to facilitate their practical application in design and operation, in 1983 the IAWPRC (subsequently IAWQ, now IWA) set up a Task Group on Mathematical Modelling for Design and Operation of Biological Wastewater Treatment. The Task Group proposed a general kinetic model for organic matter and nitrogenous material removal, called activated sludge model no. 1, ASM1. ASM1, as it was introduced in 1987 (Henze et al., 1987), has become a major reference for many scientific and practical projects. Today, mathematical models related to ASM1 are implemented in various computer codes for the simulation of the behavior of activated sludge systems treating domestic wastewater. ASM1 can predict oxygen demand, sludge production, nitrification and denitrification for activated sludge systems, i.e., COD and N removal from the wastewater. In summary, ASM1 includes growth and death processes for two groups of organisms: (1) ordinary heterotrophic organisms (OHOs) and (2) autotrophic nitrifying organisms (ANOs), and the hydrolysis of particulate slowly biodegradable organics.

The IWA Task Group extended ASM1 to include biological excess P removal (BEPR), to form ASM2 (Henze et al., 1995; Hu et al., 2003). Considering ASM1 and ASM2 model defects, the Task Group has decided to propose the Activated Sludge Model No. 3 (ASM3) which should correct for all these defects and which could become a standard again. ASM3 relates to the same dominating phenomena as does ASM1: oxygen consumption, sludge production, nitrification and denitrification in activated sludge systems treating primarily domestic wastewater. Biological phosphorus removal is contained in the Activated Sludge Model No.2 (Henze et al., 1995) and will not be considered in ASM3.

The following components are used in ASM3; concentrations of soluble components are characterized by S_* and particulate components by X_*. Within the activated sludge systems, the particulate components are assumed to be associated with the activated sludge (flocculated onto the activated sludge or contained within the active biomass). Particulate components can be concentrated by sedimentation/thickening in clarifiers, whereas soluble components can only be transported with water. Only soluble components may carry an ionic charge. One important difference relative to ASM1 and ASM2 is that in ASM3 soluble and particulate components can better be differentiated with filtration by over 0.45 μm membrane filters whereas a significant fraction of slowly biodegradable organic substrates, Xs in ASMI and ASM2 would be contained in the filtrate of the inflowing wastewater (Gujer et al., 1999).

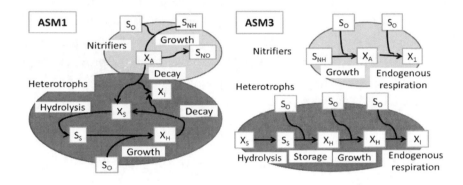

Figure 6. Flow of COD in ASM 1 and ASM3.

ASM3 includes only the microbiological transformation processes. Chemical precipitation processes are not included (Henze et al., 1995). ASM3 considers the following transformation processes: (1) Hydrolysis, (2) Aerobic storage of readily biodegradable substrate, (3) Anoxic storage of readily biodegradable substrate, (4) Aerobic growth of heterotrophs, (5) Anoxic growth of heterotrophs, (6) Aerobic endogenous respiration, (7) Anoxic endogenous respiration, (8) Aerobic respiration of storage products, and (9) Anoxic respiration of storage products. As compared with ASM1, ASM3 includes a more detailed description of internal cell processes (storage) and allows for better adjustment of decay processes to the environmental conditions. The importance of hydrolysis has been reduced and degradation of soluble and particulate organic nitrogen has been integrated into the hydrolysis, decay and growth process.

The flow of COD in ASM1 is rather complex. The death (decay) regeneration cycle of the heterotrophs and the decay process of nitrifiers are strongly interrelated (Figure 6). The two decay processes differ significantly in their details. This results in differing and confusing meanings of the two decay rates in ASM1. In ASM3, all the conversion processes of the two groups of organisms are clearly separated and decay processes are described with identical models (Figure 6) (Gujer et al., 1999). Similar to ASM2, ASM3 includes internal cell storage compounds. This requires the biomass to be modelled with the cell's internal structure. Decay processes must include both fractions of the biomass, hence four decay processes are required (aerobic and anoxic loss of X_H as well as X_{STO}) and the kinetics of the growth processes (aerobic and anoxic) must relate to the ratio of X_{STO}/X_H. ASM3 has not yet been tested against a large variety of experimental data. It is expected that future improvements of model structure may still be required, especially for the description of the storage phenomena.

5.3. The Effect of Toxic Carbon Source on the Reaction of Activated Sludge

The measurement of EPS, SMP and X_{STO} is one way to understand the impact of toxic carbon on the reaction of the activated sludge. In order to deeply understand the mechanism and the specific reaction process, however, the measurement of the three components is insufficient. As mentioned above, the mathematical model is an important tool to help us understand the metabolic process of the activated sludge. For example, based on ASM3, Laspidou et al. (2002a,b) and Ni et al. (2009) have built and developed a unified model to describe the variations of EPS, SMP and X_{STO} during the reaction process (Figure 7). The model can be applicable for modeling all three microbial products (EPS, SMP, and X_{STO}) in activated sludge under feast-famine conditions.

A mathematical model has also been used as the tool to investigate the effect of toxic carbon source (phenol) on the reaction of the activated sludge. The relationships of EPS, SMP, and X_{STO}, the effects of the toxic organics on the death of the activated sludge cells and the carbon flows can be interpreted. Wu et al. (2018) have developed an integrated model based on some previous studies (Laspidou et al., 2002a,b; Ni et al., 2009). The carbon flows of the integrated model are shown in Figure 8. In the previous activated sludge models, the reaction process is divided into two periods, external substrates degradation period (feat period) and endogenous respiration period (famine period).

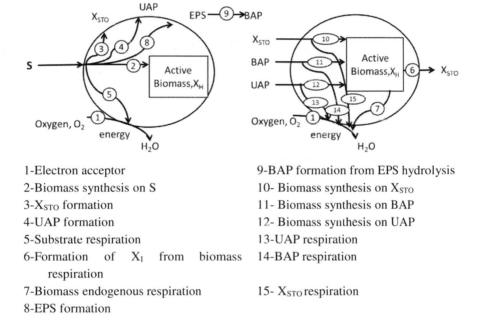

1-Electron acceptor
2-Biomass synthesis on S
3-X_{STO} formation
4-UAP formation
5-Substrate respiration
6-Formation of X_I from biomass respiration
7-Biomass endogenous respiration
8-EPS formation

9-BAP formation from EPS hydrolysis
10- Biomass synthesis on X_{STO}
11- Biomass synthesis on BAP
12- Biomass synthesis on UAP
13-UAP respiration
14-BAP respiration
15- X_{STO} respiration

Figure 7. Schematic diagram of electron flows from the external substrate (left side) and, regarding active and inert biomass, EPS, SMP, and X_{STO} (right side) for the expanded unified model (Ni et al., 2009).

The integrated model developed by Wu et al. (2018) two distinct metabolic periods. All 19 reaction processes can occur simultaneously during the aeration period. According to the previous studies, the toxic substrates can increase the death rate of microbial cells and then increase the concentrations of EPS and SMP (Zhang et al., 2010; Han et al., 2013; Wu et al., 2016). The new EPS formation pathway, the lysis of active biomass, is introduced into the integrated model (reaction 19). In addition, the biomass synthesis on EPS and SMP respiration pathways (reactions 11,

12 and 17) are also added into the metabolic model. The integrated model can be used to predict the variations of three microbial products (EPS, SMP, and X_{STO}) in activated sludge in aeration conditions.

EPS is the viscous skeleton of the activated sludge flocs (Sun et al., 2016). According to the previous studies (Wu et al., 2016; Laspidou et al., 2002b), the kinetics of EPS have been established. Equation 2 (Table 2) is the mass balance, including the formation and loss of EPS in the batch system.

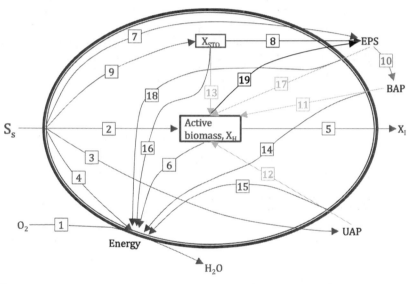

1-Electron acceptor

2-Biomass synthesis on S

3-UAP formation on S

4-Substrate respiration

5-Formation of X_I from biomass respiration

6-Biomass endogenous respiration

7-EPS formation on S

8-EPS formation on X_{STO}

9-X_{STO} formation on S

10-BAP formation from EPS hydrolysis

11-Biomass synthesis on BAP

12-Biomass synthesis on UAP

13-Biomass synthesis on X_{STO}

14-BAP respiration

15-UAP respiration

16-X_{STO} respiration

17-Biomass synthesis on EPS

18-EPS respiration

19-EPS formation from death cells

Figure 8. Schematic diagram of electron flows of the integrated model (Wu et al. 2018).

Table 2. The mass balance equations of the integrated model (Wu et al. 2018)

Component	Equation	Equation No.
S	$$\frac{dS_S(t)}{dt} = -\frac{\mu_H}{Y_H}M_S(t)M_O(t)X_H(t)$$	(1)
EPS	$$\frac{dX_{EPS}(t)}{dt} = k_{EPS}\frac{\mu_H}{Y_H}M_S(t)M_O(t)X_H(t) + k_{STO,EPS}\frac{\mu_{H,STO}}{Y_{H,STO}}I_S(t)M_O(t)M_{STO}(t)X_{STO}(t) -$$ $$k_{decay}X_{EPS}(t) - k_{hyd}X_{EPS}(t) + k_{death}X_H(t) - \frac{\mu_{EPS}}{Y_{EPS}}M_{EPS}(t)M_O(t)X_H(t)$$	(2)
UAP	$$\frac{dS_{UAP}(t)}{dt} = k_{UAP}\frac{\mu_H}{Y_H}M_S(t)M_O(t)X_H(t) - \frac{\mu_{UAP}}{Y_{UAP}}M_{UAP}(t)M_O(t)X_H(t)$$	(3)
BAP	$$\frac{dS_{BAP}(t)}{dt} = k_{hyd}X_{EPS}(t) - \frac{\mu_{BAP}}{Y_{BAP}}M_{BAP}(t)M_O(t)X_H(t)$$	(4)
X_{STO}	$$\frac{dX_{STO}(t)}{dt} = k_{STO}\frac{\mu_H}{Y_H}M_S(t)M_O(t)X_H(t) - \frac{\mu_{H,STO}}{Y_{H,STO}}M_{STO}(t)I_S(t)M_O(t)X_H(t) -$$ $$k_{STO,EPS}\,\mu_{H,STO}M_{STO}(t)I_S(t)M_O(t)M_{STO}(t)X_{STO}(t)$$	(5)

Table 2. (Continued)

Component	Equation	Equation No.
X_H	$\frac{dX_H(t)}{dt} = (1 - k_{EPS} - k_{UAP} - k_{STO})\mu_H M_S(t)M_O(t)X_H(t) +$ $\mu_{H,STO}M_{STO}(t)I_S(t)M_O(t)X_H(t) + \mu_{UAP}M_{UAP}(t)M_O(t)X_H(t) +$ $\mu_{BAP}M_{BAP}(t)M_O(t)X_H(t) - b_H M_O(t)X_H(t) +$ $\mu_{EPS}M_{EPS}(t)M_O(t)X_H(t) - k_{death}X_H(t) - m_{H,S}M_S(t)M_O(t)X_H(t) -$ $m_{H,STO}M_{STO}(t)M_S(t)M_O(t)X_H(t)$	(6)
X_I	$\frac{dX_I(t)}{dt} = -f_I b_H M_O(t)X_H(t)$	(7)
DO	$\frac{dS_O(t)}{dt} = k_{La}(S_O^* - S_O(t)) -$ $[1 - k_{EPS} - k_{UAP} - k_{STO} - Y_H(1 - k_{EPS} - k_{UAP} - k_{STO})]\mu_H M_S(t)M_O(t)X_H(t) -$ $\frac{1 - Y_{H,STO}}{Y_{H,STO}}\mu_{H,STO}M_{STO}(t)I_S(t)M_O(t)X_H - (1 - f_I)b_H M_O(t)X_H(t) -$ $\frac{1 - Y_{UAP}}{Y_{UAP}}\mu_{UAP}M_{UAP}(t)M_O(t)X_H(t) - \frac{1 - Y_{BAP}}{Y_{BAP}}\mu_{BAP}M_{BAP}(t)M_O(t)X_H(t) -$ $\frac{1 - Y_{H,STO}}{Y_{H,STO}}k_{STO,EPS}\mu_{H,STO}M_{STO}(t)I_S(t)M_O(t)X_{STO}(t) - \frac{1 - Y_{EPS}}{Y_{EPS}}\mu_{EPS}M_{EPS}(t)M_O(t)X_H(t) -$ $(m_{H,S} + m_{H,STO}M_{STO}(t))M_S(t)M_O(t)X_H(t)$	(8)

Table 3. Monod kinetic functions in the integrated model
(Wu et al. 2018)

Kinetic expressions	Kinetics definitions
$M_S(t) = \dfrac{S_S(t)}{K_S + S_S(t)}$	Monod kinetic equation of exogenous substrates S_S
$M_O(t) = \dfrac{S_O(t)}{K_O + S_O(t)}$	Monod kinetic equation of DO
$M_{UAP}(t) = \dfrac{S_{UAP}(t)}{K_{UAP} + S_{UAP}(t)}$	Monod kinetic equation of UAP
$M_{BAP}(t) = \dfrac{S_{BAP}(t)}{K_{BAP} + S_{BAP}(t)}$	Monod kinetic equation of BAP
$M_{STO}(t) = \dfrac{X_{STO}(t)/X_H(t)}{K_{STO} + X_{STO}(t)/X_H(t)}$	Surface saturation kinetics of X_{STO} degradation
$I_S(t) = \dfrac{K_S}{K_S + S_S(t)}$	Monod kinetics of inhibition item of exogenous substrates S_S
$M_{EPS}(t) = \dfrac{X_{EPS}(t)/X_H(t)}{K_{EPS} + X_{EPS}(t)/X_H(t)}$	Monod kinetic equation of EPS

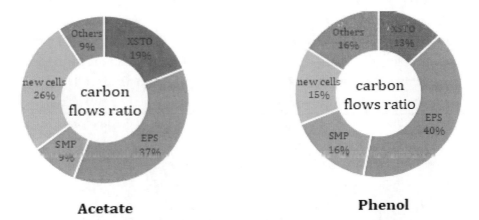

Figure 8. The electron flows ratios of different carbon sources in the steady state.

Compared to previous studies, they introduce the formation of EPS from the reaction of X_{STO} (Reaction 8). In addition, to describe the effect of toxic substances, Wu et al. (2018) also added other pathways of the EPS formation, as shown in Figure 8. For example, $k_{death}X_H$ is the production of

EPS from the death of activated sludge. The transformation of EPS and BAP is the same as Ni et al. (2009) and the hydrolysis rate is proportional to the EPS concentration (Wu et al., 2016). Equations 3 and 4 describe the mass balance of UAP and BAP in the batch reactor, as the SMP can be divided into these two categories (Ni et al., 2012). In this model, the formation of SMP is the same as Ni et al. (2009) and it will not be discussed in detail. Equation 5 is the mass balance of X_{STO}, which describes the formation and utilization of X_{STO} in the batch reactor. The simultaneous storage and growth concept is the base of X_{STO} transformation (Laspidou et al., 2002b; Ni et al., 2009). Compared to previous studies, EPS formation on X_{STO} has been added into the integrated model as mentioned above (Reaction 8). The kinetics of X_I is the same as the previous study (Ni et al., 2009). However, due to the consideration of toxic effect on reaction and the transformation of EPS and X_{STO}, the kinetics for oxygen transfer and consumption are more complex than Ni et al. (2009), as shown in equation 8. The equivalent amount of oxygen demand, which is proportional to electron equivalents (8 g of O_2 per e-equivalent), is the general unit for the calculation of all components.

Based on the electron transfer processes and pathways, the mass balance equations of the model are proposed (Table 2). The kinetic equations are the non-steady-state mass balance equations of the 8 components. As mentioned above, the model has no distinct feast and famine periods. The Monod kinetic equations, which play a role of switching functions, are introduced into the mass balance equations (Table 3). Take equation 3 for example, $k_{UPA}(\mu_H/Y_H)M_S(t)M_O(t)X_H(t)$ is the formation of UAP based on substrates. When the substrates concentration is high in the initial aeration period, the formation rate, controlled by $M_S(t)$ = $S_S(t)/[K_S + S_S(t)]$, is also high. When the exogenous substrates concentration is low, the formation rate of UAP, controlled by $M_S(t)$ is closed to 0, and then UAP turns to be consumed as the carbon source. The switching Monod kinetic equations can control the dynamic variation of different components during the aeration time. All the simulations can be performed by the MATLAB software package.

The stoichiometric and kinetic coefficients of the model can be checked in Wu et al. (2018). The integrated model was calibrated and validated by the experimental data and it was found that the model simulations matched all experimental measurements. In the steady state, the toxicity of phenol can result in higher microbial cells death rate (0.1637 h^{-1} vs 0.0028 h^{-1}) and a higher decay rate coefficient of biomass (0.0115 h^{-1} vs 0.0107 h^{-1}) than acetate. In addition, the utilization-associated products (UAP) and extracellular polymeric substances (EPS) formation coefficients of phenol are higher than that of acetate, indicating that more carbon flows into the extracellular components, such as soluble microbial products (SMP), when degrading toxic organics. In the non-steady state of feeding phenol, the yield coefficient for growth and maximum specific growth rate are very low in the first few days (1-10 d), while the decay rate coefficient of biomass and microbial cells death rate are relatively high. The carbon flows of the activated sludge feeding with different carbon sources can be calculated by the integrated model (Figure 8). The model provides insights into the difference of the dynamic reaction with different carbon sources in the batch reactor. When feeding acetate in the steady state, on average, 19% of the carbon could be transferred into X_{STO}, 37% of the carbon could be transferred into EPS, 9% of the carbon could be transferred into SMP, 26% of the carbon could be transferred into new cells and 9% of the carbon was oxidized by the O_2 and transferred into others. While feeding phenol, 13% of the carbon could be transferred into X_{STO}, 40% of the carbon could be transferred into EPS, 16% of the carbon could be transferred into SMP, 15% of the carbon could be transferred into new cells and 16% of the carbon was oxidized by the O_2 and transferred into others. In summary, when feeding toxic phenol, less new cell growth and higher EPS and SMP production were conducted compared to using acetate as the carbon source. The statistical analysis of different components were performed ($0.001 < p = 0.004$-$0.009 < 0.01$ for X_{STO}, SMP, new cells and others while $p = 0.054$ for EPS) and it was found that the carbon flows during acetate and phenol removal were significantly different. The toxic carbon source can cause the reaction variation of the activated sludge, resulting in higher residual effluent dissolved organic carbon such as SMP than an easily biodegraded carbon source. In addition, the toxicity of carbon can weaken the cells' assimilation.

REFERENCES

Allen, M. S., Welch, K. T., Prebyl, B. S., Baker, D. C., Meyers, A. J., Sayler, G. S. (2004). Analysis and glycosyl composition of the exopolysaccharide isolated from the floc-forming wastewater bacterium *Thauera* sp. MZ1T. *Environ. Microbiol.*, 6, 780–790.

Aquino, S. F., Stuckey, D. C. (2006). Chromatographic characterization of dissolved organics in effluents from two anaerobic reactors treating synthetic wastewater. *Water Sci. Technol.*, 54 (2), 193–198.

Aquino, S. F., Stuckey, D. C. (2004). Soluble microbial products formation in anaerobic chemostats in the presence of toxic compounds. *Water Res.*, 38, 255–266.

Avella, A. C., GÖner, T., de Donato, P. (2010). The pitfalls of protein quantification in wastewater treatment studies. *Sci. Total. Environ.*, 408, 4906–4909.

Barker, D. J., Stuckey, D. C. (1999). A review of soluble microbial products (SMP) in wastewater treatment systems. *Water Res.*, 33, 3063–3082.

Beech, I. B., Cheung, C. W. S., Johnsom, D. B., Smith, J. R. (1996). Comparative studies of bacterial biofilms on steel surface using atomic force microscopy and environmental scanning electron microscopy. *Biofouling*, 10(1-3), 65–77.

Comte, S., Guibaud, G., Baudu, M. (2006). Relations between extraction protocols for activated sludge extracellular polymeric substances (EPS) and EPS complexation properties: Part I. Comparison of the efficiency of eight EPS extraction methods. *Enzyme Microb. Technol.*, 38, 237–245.

Comte, S., Guibaud, G., Baudu, M. (2007). Effect of extraction method on EPS from activated sludge: an HPSEC investigation. *J. Hazard. Mater.*, 140, 129–137.

Dignac, M. F., Urbain, V., Rybacki, D., Bruchet, A, Snidaro, D, Scribe, P. (1998). Chemical description of extracellular polymeric substances: implication on activated sludge floc structure. *Water Sci. Technol.,* 38(8–9), 45–53.

Dionisi, D., Majone, M., Bellani, A., Viggi Cruz, C., Beccari, M. (2008). Role of biomass adaptation in the removal of formic acid in sequencing batch reactors. *Water Sci. Technol.,* 58 (2), 303-307.

Dufrene, Y. F., Rouxhet, P. G. (1996). X-ray photoelectron spectroscopy analysis of the surface composition of *Azospirillum brasilense* in relation to growth conditions. *Colloids Surf. B,* 7, 271–279.

Frolund, B., Palmgren R., Keiding K., Nielsen P. H. (1996). Extraction of extracellular polymers from activated sludge using a cation exchange resin. *Water Res.,* 30, 1749–58.

Frolund, B., Griebe T., Nielsen P. H. (1995). Enzymatic activity in the activated-sludge floc matrix. *Appl. Microbiol. Biotechnol.,* 43, 755–761.

Gujer, W., Henze, M., Mino, T., Loosdrecht, Mv. (1999). Activated sludge model No. 3 *Water Sci. Tech.,* 39(1), 183–199.

Han, J., Liu, Y., Liu, X., Zhang, Y., Yan, Y., Dai, R., Zha, X., Wang, C. (2013). The effect of continuous Zn (II) exposure on the organic degradation capability and soluble microbial products (SMP) of activated sludge. *J. Hazard. Mater.,* 244–245, 489–494.

Henriques, I. D. S., Love, N. G. (2007). The role of extracellular polymeric substances in the toxicity response of activated sludge bacteria to chemical toxins. *Water Res.,* 41, 4177–4185.

Henze, M., Grady, C. P. L. Jr., Gujer, W., Marais, Gv. R., Matsuo, T. (1987). Activated sludge model no. 1. *IAWQ Scientific and Technical Report No.1,* IAWQ, London, 33pp.

Henze, M., Gujer, W., Mino, T., Matsuo, T., Wentzel, M. C., Marais, Gv. R. (1995). Activated sludge model No. 2 *IAWQ Scientific and Technical Report No. 3,* IAWQ, London, UK, 32pp.

Hu, Z. R., S.otemann, S. W., Moodley, R., Wentzel, M. C., Ekama, G. A. (2003). Experimental investigation on the external nitrification biological nutrient removal activated sludge system—The ENBNRAS system. *Biotech. Bioeng.*, 83(3), 260–273.

Huang, G., Jin, G., Wu, J., Liu, Y. (2008). Effects of glucose and phenol on soluble microbial products (SMP) in sequencing batch reactor systems. *Int. Biodeterior. Biodegrad.*, 62, 104–108.

Imai, A., Fukushima, T., Matsushige, K., Kim, Y. H. Choi, K. (2002). Characterization of dissolved organic matter in effluents from wastewater treatment plants. *Water Res.*, 36, 859–870.

Ji, L., Zhou, J. (2006). Influence of aeration on microbial polymers and membrane fouling in submerged membrane bioreactors. *J. Membr. Sci.*, 276, 168–177.

Jin, B., Wilén, B. M., Lant, P. (2004). Impacts of morphological, physical and chemical properties of sludge flocs on dewaterability of activated sludge. *Chem. Eng. J.*, 98, 115–126.

Jørgensen, M. K., Nierychlo, M., Nielsen, A. H., Larsen, P., Christensen, M. L., Nielsen, P. H. (2017). Unified understanding of physico-chemical properties of activated sludge and fouling propensity. *Water Res.*, 120, 117–132.

Laspidou, C. S., Rittmann, B. E. (2002a). Unified theory for extracellular polymeric substances, soluble microbial products, and active and inert biomass. *Water Res.*, 36, 2711–2720.

Laspidou, C. S., Rittmann, B. E. (2002b). Non-steady state modeling of extracellular polymeric substances, soluble microbial products, and active and inert biomass. *Water Res.*, 36, 1983–1992.

Lattner, D, Flemming, H. C, Mayer, C. (2003). 13C-NMR study of the interaction of bacterial alginate with bivalent cations. *Int. J. Biol. Macromol.*, 31, 81–88.

Liu, H., Fang, H. H. P. (2002a). Extraction of extracellular polymeric substances (EPS) of sludges. *J. Biotechnol.*, 95, 249–256.

Li, X., Logan, B. (2004). Analysis of bacterial adhesion using a gradient force analysis method and colloid probe atomic force microscopy. *Langmuir*, 20, 8817–8822.

Liu, X. M., Sheng, G. P., Luo, H. W., Zhang, F., Yuan, S. J., Xu, J., Zeng, R. J., Wu, J. G., Yu, H. Q. (2010). Contribution of extracellular polymeric substances (EPS) to the sludge aggregation. *Environ. Sci. Technol.*, 44 (11), 4355–4360.

Manca, M. C., Lama, L., Improta, R., Esposito, E., Gambacorta, A., Nicolaus, B. (1996). Chemical composition of two exopolysaccharides from *Bacillus thermoantarcticus*. *Appl. Environ. Microbiol.*, 62, 3265–3269.

Mehrez, R., Ernst, M., Jekel, M. (2007). Development of a continuous protein and polysaccharide measurement method by sequential injection analysis for application in membrane bioreactor systems. *Water Sci. Technol.*, 56 (6), 163–171.

Ni, B. J., Fang, F., Rittmann, B. E., Yu, H. Q. (2009). Modeling microbial products in activated sludge under feast-famine conditions. *Environ. Sci. Technol.* 43, 2489–2497.

Ni, B. J., Yu, H. Q. (2012). Microbial products of activated sludge in biological wastewater treatment systems: a critical review. *Crit. Rev. Environ. Sci. Technol.*, 42,187–223.

Nielsen P. H., Jahn A. (1999). Extraction of EPS. In: Wingender J., Neu T. R., Flemming H. C., editors. *Microbial extracellular polymeric substances: characterization, structure and function.* Berlin Heidelberg: Springer-Verlag; p. 49–72. Chapter 3.

Oehmen, A., Yuan, Z., Blackall, L. L., Keller, J. (2005). Comparison of acetate and propionate uptake by polyphosphate accumulating organisms and glycogen accumulating organisms. *Biotechnol. Bioeng.*, 91, 162–168.

Omoike, A., Chorover, J. (2004) Spectroscopic study of extracellular polymeric substances from *Bacillus subtilis*: aqueous chemistry and adsorption effects. *Biomacromolecules*, 5, 1219–1230.

Park, C., Novak, J. T. (2007). Characterization of activated sludge exocellular polymers using several cation-associated extraction methods. *Water Res.*, 41, 1679–1688.

Potvin, C. M., Zhou, H. (2010). Interference by the activated sludge matrix on the analysis of soluble microbial products in wastewater. *Chemosphere*, 85, 1139–1145.

Rosenberger, S., Evenblij, H., te Poele, S., Wintgens, T., Laabs, C. (2005). The importance of liquid phase analyses to understand fouling in membrane assisted activated sludge processes—six case studies of different European research groups. *J. Membr. Sci.*, 263, 113–126.

Sheng, G. P., Yu, H. Q., Wang, C. M. (2006b). FTIR-spectral analysis of two photosynthetic hydrogen producing strains and their extracellular polymeric substances. *Appl. Microbiol. Biotechnol.*, 73, 204–210.

Sheng, G. P., Yu, H. Q., Li, X. Y. (2006a). Stability of sludge flocs under shear conditions: roles of extracellular polymeric substances (EPS). *Biotechnol. Bioeng.*, 93, 1095–1102.

Sheng, G. P., Yu, H. Q., Li, X. Y. (2010). Extracellular polymeric substances (EPS) of microbial aggregates in biological wastewater treatment systems: A review. *Biotechnol. Adv.*, 28, 882–894.

Sheng, G. P., Yu, H. Q., Yu, Z. (2005a). Extraction of the extracellular polymeric substances from a photosynthetic bacterium *Rhodopseudomonas acidophila*. *Appl. Microbiol. Biotechnol.*, 67, 125–130.

Shi, Y., Huang, J., Zeng, G., Gu, Y., Chen, Y., Hu, Y., Tang, B., Zhou, J., Yang, Y., Shi, L. (2017). Exploiting extracellular polymeric substances (EPS) controlling strategies for performance enhancement of biological wastewater treatments: an overview. *Chemosphere*, 180, 396–411.

Staudt, C., Horn, H., Hempel, D. C., Neu, T. R. (2004). Volumetric measurements of bacterial cells and extracellular polymeric substance glycoconjugates in biofilms. *Biotechnol. Bioeng.*, 88, 585–592.

Sun, J., Guo, L., Li, Q., Zhao, Y., Gao, M., She, Z., Wang, G. (2016). Structural and functional properties of organic matters in extracellular polymeric substances (EPS) and dissolved organic matters (DOM) after heat pretreatment with waste sludge. *Bioresource. Technol.*, 219, 614-623.

Tan, T. W., Ng, H. Y. (2008). Influence of mixed liquor recycle ratio and dissolved oxygen on performance of pre-denitrification submerged membrane bioreactors. *Water Res.*, 42, 1122–1132.

van der Aa, B. C., Dufrene, Y. F. (2002). In situ characterization of bacterial extracellular polymeric substances by AFM. *Colloids Surf. B,* 23, 173–182.

van Loosdrecht, M. C. M., Pot, M., Heijnen, J. (1997). Importance of bacterial storage polymers in bioprocesses. *Water Res.,* 35, 41–47.

Wingender J., Neu T. R., Flemming H. C. (1999). What are bacterial extracellular polymeric substances? In: Wingender J, Neu TR, Flemming HC, editors. *Microbial extracellular polymeric substances: characterization, structures and function.* Berlin Heidelberg: Springer-Verlag; p. 1-18. Chapter 1.

Wu, C., Zhou, Y., Song, J. (2016). The activated sludge metabolic characteristics changing sole carbon source from readily biodegradable acetate to toxic phenol. *Water Sci. Technol.,* 73(10), 2324-2331.

Wu, C., Zhou, Y., Zhang, S., Xu, M., Song, J. (2018). The effect of toxic carbon source on the reaction of activated sludge in the batch reactor. *Chemosphere,* 194, 784-792.

Wu, W., Duan, T., Song, H., Li, Y., Yu, A., Zhang, L., Li, A. (2015). The effect of continuous Ni(II) exposure on the organic degradation and soluble microbial product (SMP) formation in two-phase anaerobic reactor. *J. Environ. Sci.,* 33, 78-87.

Yu, G. H., He, P. J., Shao, L. M. (2009b). Characteristics of extracellular polymeric substances (EPS) fractions from excess sludges and their effects on bioflocculability. *Bioresource. Technol.,* 100, 3193–3198.

Yu, G., He, P., Shao, L., Zhu, Y. (2009a). Enzyme extraction by ultrasound from sludge flocs. *J. Environ. Sci.,* 21, 204–210.

Yan, Y., Wang, Y., Liu, Y., Liu, X., Yao, C., Ma, L. (2015). Effect of continuously dosing Cu(II) on pollutant removal and soluble microbial products in a sequencing batch reactor. *Water Sci.Technol.,* 72 (9), 1653–1661.

Zhang, Z. P., Zhang, T. (2010). Characterization of soluble microbial products (SMP) under stressful conditions. *Water Res.,* 44, 5499–5509.

Zhang, T., Fang, H. H. P. (2001). Quantification of extracellular poplymeric substances in biofilms by confocal laser scanning microscopy. *Biotechnol. Lett.,* 23, 405–409.

substrates that are most beneficial of xenobiotic degradation rate. 2,4-dichlorophenol acid (2,4-D) was used representative xenobiotic organic compounds, while peptone and sugar used for auxiliary substrates. The activated sludge was completely break down 100mg/l of 2,4-D for three consecutive times. The different concentrations between biogenic substracts of sucrose and peptone were fed separately or combined into the medium containing 200mg/l of 2,4-D and 140mg SS/l of activated sludge. The results showed that sugar and peptone could affect 2,4-D degradation rate to several different degree at different concentrations. In separate supplementation, 2,4-D degradation completed within 25 hours, 40mg/l sugar and 150mg/l peptone concentrations were found to be the optimal concentrations. In combined case, 2,4-D was consumed totally within 20 hours and the optimal concentration of the combined sugar and peptone concentrations were 40 and 150mg/l, respectively.

1. INTRODUCTION

Xenobiotic organic compounds such as phenoxy acid herbicides are foreign to most indigenous microorganisms, including those in activated sludge. In addition to their hard-to-treat nature, xenobiotics can be uncoupled of the microorganisms' metabolism of biogenic substrates. The uncoupling, or the energy-spilling reactions, occurs when catabolism energy is dissipated or diverted away from anabolism through the metabolism of a substrate in bacterial cells [1]. Although xenobiotics containing wastewater can be suitably treated using biological methods [2, 3, 4, 5], the xenobiotic nature of the pollutant requires the treatment plant microorganisms (typically activated sludge) to go through an acclimation phase before the microorganisms evolve the degradation capability for treating the influent xenobiotics. Thus, biodegradation method is used widely to treat xenobiotics in water treatment methods. One of the popular xenobiotics is 2,4-Dichlorophenoxyacetic acid (2,4-D), a common joint systemic herbicide used in the control of broadleaf weeds. Therefore, the target xenobiotic organism pollutant was 2,4-D and activated sludge was an agent treatment in this study. The activated sludge process reduced or eliminated wastewater toxicity from a variety of sources along a wastewater collection system [6, 7]. Eckenfielder [8] suggested that

In: The Activated Sludge Process
Editor: Benjamin Lefèbvre

ISBN: 978-1-53615-202-9
© 2019 Nova Science Publishers, Inc.

Chapter 6

THE EFFECTS OF COMBINED GROWTH OF BIOGENIC AND XENOBIOTIC SUBSTRATES ON DEGRADATION OF XENOBIOTIC BY ACTIVATED SLUDGE

Nguyen Phuc Thien[1,2,*]*, PhD, Le Quoc Tuan*[1]*, PhD and Doan Quang Tri*[3]*, PhD*

[1]Nguyen Tat Thanh University, Hồ Chí Minh, Vietnam
[2]Ho Chi Minh City University of Technology -
Vietnam National University HCM, Ho Chi Minh, Vietnam
[3]National Center for Hydrometeorological Forecasting (NCHMF),
Ha Noi, Vietnam

ABSTRACT

The purpose of this study was to research about supplementation of different concentrations of the substrate on the degradation rate of xenobiotic and to determine the optimal concentrations of the auxiliary

* Corresponding Author Email: thien451986@gmail.com.

aquatic toxicity data on specific organic chemicals should not be applied to industrial wastewater effluents because many of these compounds were removed during biological treatment. Several authors [9, 10] suggested that the fed-batch technique could evaluate toxicity to activated sludge with substrate continuously added to a batch reactor and accumulation of substrate followed with time. Toxic or inhibitory substrates exhibited an upward slope in the substrate accumulation curve, and nonlinear curve fitting techniques were used to determine inhibition constants.

Activated sludge biomass grown on the feed of biogenic substrate must be in a healthy physiological condition [11]. However, inconsistent results are found in literature about the effects of biogenic organics on the degradation of man-made xenobiotics or hydrocarbons. There have been some study cases of both beneficial and adverse effects of biogenic organics on xenobiotic degradation. The beneficial good cases include: citrate on toluene [12]; natural amino acids on mono-substituted phenol [13]; natural organics such as manure on dichloro two chloros and a nitro-herbicides [14]; fatty acids on soil hydrocarbon [15]; pyruvate on naphthalene [16]. Conversely, adverse cases are also found: glucose or amino acids in on xylene and toluene [17]; ethanol in on benzene, toluene and xylene [18]; glycolic acid and glucose on p-cresol [19]; yeast extract and milk on 3-nitrobenzoate, 4-chlorobenzoate, 4-chlorophenol [20]. Diauxic growth is a phenomenon occurs when microorganism is grown on cultural medium with the presence of two substrates, one of which is easier for the microorganism to metabolize. Sugar is consumed first, and it leads to a rapid growth. Only after the easier more natural substrate has been exhausted, the cells switch to the second, resulting in two separate growth phases. There are some previous studies which the effects of diauxic growth on 2,4-D acclimation and degradation [21]. The results showed that: If we feed sugar with a proper concentration and at the proper time, it can make 2,4-D acclimation lag time shorten. On the contrary, it can make the acclimation lag time longer. This is explained that sugar is an easier substrate than 2,4-D for the microbial cell to utilize, and thus the biomass temporarily ignores or escapes from the xenobiotic acclimation stress. This pause in of the process lets the xenobiotic still be intact until sugar is consumed entirely completely, thereby causing the elongation of the acclimation lag time. The purpose of this study, therefore, was to determine effects of auxiliary substrates on 2,4-D degradation and detail profiles of ATP that contained in activated sludge cells during the cells' degradation of a model xenobiotic compound 2,4- dichloro-phenoxyacetic acid (2,4-D). Sugar and peptone were supplemented with into cultural medium contained 2,4-D to determine the optimal concentrations of each the auxiliary substrates in separate or double. Generally, sugar and peptone are used as primary major carbon and nitrogen sources in a medium of microbial cultivation. However, 2,4-D plays a role as a carbon source in this study. Thus, sugar and peptone were used as the representation of the auxiliary substrates. In this study case, sugar sucrose and feed of 2,4-D, peptone and feed of 2,4-D with peptone are fed transferred into the cultural medium at the same time. Therefore, we can find out which concentration is the most effective on 2,4-D degradation. Results of this study can supplement for the knowledge about 2,4-D biodegradation, and apply to actual real water treatment system. The efficiency and applicability consist of economic efficiency when the methods are applying to actual xenobiotic treatment systems. However, this study can be used as references for studies of relative compounds and other xenobiotic organics.

2. MATERIALS AND METHODS

As mentioned above, 2,4-D was used to be target substrate, and activated sludge was used to be agent treatment. Peptone and sugar are considered as auxiliary substrates that could affect 2,4-D degradation by activated sludge.

2.1. Suspended Solid (SS) Measurements

Suspended solid is considered as a representation of microbe biomass in samples. It is measured one time per day during 2,4-D degradation in this study. The procedure is used, including the following steps: 1. The filter papers (pore size 0.45 μm) are washed by using distilled water and vacuum filtration; 2. After washing, the papers are dried about 4-5 hours at 105^0C in a drying cabinet; 3. Scaling the dried the papers before using to filter samples, getting paper weight (M_0 - g); 4. Putting V (ml) sample on the dried filter paper and filtering by vacuum filtration to separate the solid from the liquid; 5. Drying again about 4- 5 hours at 105^0C in the drying cabinet; 6. Scaling again the dried sample to get paper and SS weight (M_1 - g); 7. Calculating SS (mg/l) is showed in Equation 1:

$$SS = \frac{(M_1 - M_0)x100}{0.001\,xV}$$

(1)

where: SS is suspended solid weight (mg/l); M_0 is the weight of the initial dried paper (g); M_1 is a weight of the dried paper after filtering sample (g); V is a volume of the sample (ml).

2.2. Experimental Method to Obtain Acclimated Activated Sludge

Activated sludge was fed in separate mediums corresponding to different periods as follows: The first period was to cultivate activated sludge in a sequence batch reactor with supplementing daily sucrose 100mg/l, peptone 18mg/l, $FeCl_3$ 1mg/l, NH_4Cl 30mg/l, K_2HPO_4 200mg/l, KH_2PO_4 156mg/l and $MgSO_4$ 31.26mg/l. The second period was to make the activated sludge acclimate to 2,4-D. After feeding in the continuous medium, 140mg/l of the activated sludge was transferred to batch medium with 100mg/l 2,4-D as sole carbon resource until the activated sludge degraded completely 2,4-D in three times. The original activated sludge spent about seven days degrading completely 100mg/l 2,4-D in the first time. However, it only spent about two days in the second time, and one day for in the third time on this process. The obtained activated sludge was called A. A would be used major experiments with sugar and peptone supplementation.

2.3. Experimental Methods with (Sugar) Sucrose and Peptone

To determine effects of auxiliary substrates supplementation on 2,4-D degradation, the study conducted the experiments in separated and combined case. Experimental symbols: A stands for the reactor only includes A and 2,4-D; S_x stands for the reactor includes A, x mg/l Sugar and 2,4-D; P_y stands for the reactor includes A, y mg/l Peptone and 2,4-D; P_yS_x stands for the reactor includes A, y mg/l Peptone x mg/l Sugar and 2,4-D. All experiments were conducted with 200 mg/l 2,4-D and 140 mg/l A. In separated case: Each substrate was supplemented concurrently with 2,4-D into A medium with different concentrations. Supplemental sugar concentrations consisted of 20, 40, 60, 80, 100, 150mg/l, while supplemental peptone concentrations were 20, 40, 100, 150, 200, 300mg/l. In combined case: Both sugar and peptone are supplemented concurrently with 2,4-D into A medium with different couple concentrations. Supposing that: 0 stands for the best concentrations; "-"stands for the lower concentrations than the best concentrations; "+" stands for the higher concentrations than the best concentrations. From the supposition, the matrix of sugar and peptone concentrations is shown in Table 1.

Table 1. Matrix of sugar and peptone concentrations

Peptone / Sugar	-	0	+
-	- -	- 0	- +
0	0 -	0	0 +
+	+ -	+ 0	+ +

2,4-D concentration and SS were measured each 5 hours during the degradation process.

2.4. 2,4-D Degradation Rate Calculation

This study used Equation 2 to calculate 2,4-D degradation rate.

$$R = \frac{S}{t} \cdot \frac{(S_i - S_{i+n})}{(t_{i+n} - t_i)}$$

(2)

where: R stands for 2,4-D degradation rate (mg 2,4-D/l.hr); S is 2,4-D concentration; t stands for time (hour); i stands for first time mark, the point was chosen at 0hr; i+n stands for the second time mark, the point was chosen at 20hrs or 15hrs for separate cases or combined case, respectively.

3. RESULTS AND DISCUSSION

3.1. Effects of Different Sugar Concentrations on 2,4-D Degradation Rate

SS increased from 140mg/l to 270mg/l, depending on sugar concentration (figure 1a). Figure 1b showed that 2,4-D degradation completed within 25 hours. SS grew on higher sugar concentrations were always higher greater than the growth in on lower concentrations at the same time.

The reason is that sugar is easily consumed by microorganisms and releases a higher amount of energy. The energy could help the sludge grow faster. 2,4-D degradation rates were calculated, and shown in figure 1c. Figure 1c showed that the reactor included 40mg/l sugar, had the highest rate of 2,4-D degradation at 6.94mg/l.hr.

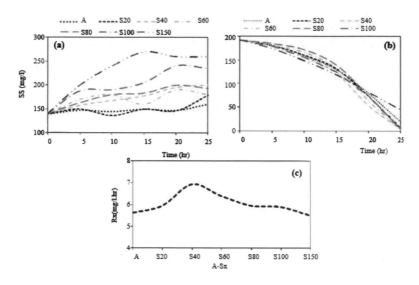

Figure 1. (a) 2,4-D degradation with sugar supplementation; (b) Activated sludge growth (SS) of sugar supplementation; (c) 2,4-D degradation rate of sugar supplementation.

For sugar concentrations that were less than 40 mg/l, the rate increased when sugar concentration increased. Because the microorganisms can easily metabolize sugar, and releases a large of energy. The energy could help the sludge grow faster. But the trend would be not similar for very high sugar concentrations such as 150mg/l. This could be explained by the diauxic growth process. Sugar is a better substrate than 2,4-D for microorganisms to utilize, and thus the biomass temporarily ignores or escapes from the xenobiotic degradation stress. This interference of the degradation lets the xenobiotic still being ignored until sugar is consumed completely. Thus, 40mg/l sugar concentration is supposed the optimal sugar concentration to enhance 2,4-D degradation rate.

3.2. Effects of Different Peptone Concentrations on 2,4-D Degradation Rate

Figure 2a showed that 2,4-D degradation completed within 25 hours. The growth of activated sludge also increased when peptone concentration increased. However, sugar made the activated sludge grow faster than peptone did at similar concentration. Therefore, sugar might release more energy than peptone did when microorganisms metabolized these substrates (figure 2b). Figure 2c showed that 2,4-D degradation rate increased when peptone concentrations were increased until peptone concentration reached higher 150mg/l. This case was not like sugar because there was a range of peptone concentrations between 150 and 200mg/l which have similar effect on 2,4-D degradation. There is been an assumption that peptone does not cause diauxic growth as sugar. Hence, the optimal peptone concentration for the best 2,4-D degradation rate was 150mg/l.

3.3. Effects of Combined Auxiliary Substrates on 2,4-D Degradation Rate

From the results obtained previously, 40mg/l sugar concentration and 150mg/l peptone concentration were supposed to be the optimal concentrations.

This section presents the combined case which used both substrates: sugar and peptone were supplemented concurrently with different couple concentrations. These different coupled concentrations were designed as shown in Table 2.

For combined sugar and peptone supplementation, 2,4-D degradation completed within about 20 hours (fugures 3a-3c); especially, the activated growth increases appreciably from 140 mg/l to 300mg/l (figure 3d). It means reasonable effects of sugar and peptone on 2,4-D degradation and the activated sludge growth are significant. The growth was obviously resulted from the auxiliary substrates, shown in figure 3.

Figure 4 showed that couple of 150mg/l peptone and 40mg/l sugar concentrations are supposed as the optimal concentrations with the rate is 8.99mg/l.hr. In this case, it is very difficult to identify a certain trend of the rate because of the combination of two auxiliary substrates. Thus, this study only determines the optimal concentrations of combined auxiliary substrates.

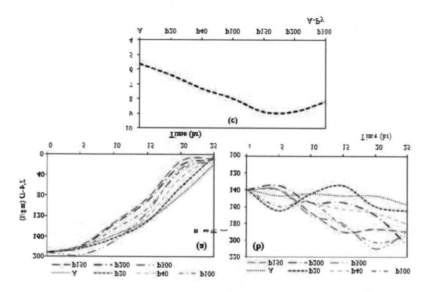

Figure 2. (a) 2,4-D degradation with peptone supplementation; (b) Activated sludge growth (SS) with peptone supplementation; (c) 2,4-D degradation rate of peptone supplementation.

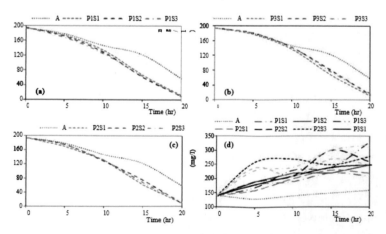

Figure 3. (a) 2,4-D degradation with sugar and peptone supplementation - P1Sx; (b) 2,4-D degradation with sugar and peptone supplementation - P3Sx; (c) 2,4-D degradation with sugar and peptone supplementation - P2Sx;(d) Activated sludge growth (SS) with sugar and peptone supplementation.

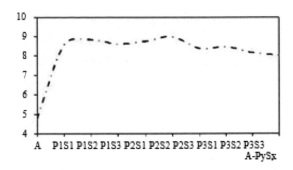

Figure 4. 2,4-D degradation rate with sugar and peptone supplementation.

Table 2. Matrix of sugar and peptone concentrations

Sugar (S) / Peptone (P)	S_1 (20 mg/l)	S_2 (40 mg/l)	S_3 (100 mg/l)
P_1 (100 mg/l)	P_1S_1	P_1S_2	P_1S_3
P_2 (150 mg/l)	P_2S_1	P_2S_2	P_2S_3
P_3 (200 mg/l)	P_3S_1	P_3S_2	P_3S_3

CONCLUSION

1. For separate supplementation, both sugar and peptone could help 2,4-D degradation rate. 2,4-D degradation completed within 25 hours. The optimal concentrations of sugar and peptone were supposed 40mg/l and 150mg/l with 6.94mg/l.hr and 150mg/l.hr of 2,4-D degradation rate, respectively. These rates were higher than the rate of un-supplemented reactor (A).
2. For combined supplementation, 2,4-D degradation completed within 20 hours. The optimal concentration of the combined sugar and peptone supplementation is 40mg/l and 150mg/l for sugar and peptone, respectively. 2,4-D degradation rate at the optimal rate was 8.99mg/l.

ACKNOWLEDGMENTS

We are thankful to Prof.Chong, Nuyk-Ming for giving us this opportunity and facilities to carry out this study. The first author would like to give special thanks to Da-Yeh University, who provides financial support for this study. A part of this research was supported by Ministry of Industry and Trade of the Socialist Republic of Vietnam under contract number: 040.2018.ĐT.BO/HĐKHCN.

REFERENCES

[1]   Russell, J. B. (2007). The energy spilling reactions of bacteria and other organisms. *Mol. Microbiol. Biotechnol.*, 13(1-3), 1-11.

[2]   Hill, N. P.; MacIntyre, A. E.; Perry, R. and Lester, J. N. (1986). Behaviour of chlorophenoxy herbicides during activated sludge treatment of municipal wastewater. *Water Res.*, 20, 45-52.

[3]   Ettala, M.; Koskela, J. and Kiesila, A. (1992). Removal of chlorophenols in a municipal sewage treatment plant using activated sludge. *Water Res.*, 26, 797-804.

[4]   Meric, S.; Eremektar, G.; Ciner, F. and Tünay, O. (2003). An OUR-based approach to determine the toxic effects of 2,4-dichloro-phenoxyacetic acid in activated sludge. *J. Hazard. Mater.*, 101 (2), 147-155.

[5]   Chin, H.; Elefsiniotis, P. and Singhal, N. (2005). An OUR-based approach to determine the toxic effects of 2,4-dichlorophenoxyacetic acid in activated sludge. *J. Environ. Eng. Sci.*, 4 (1), 57-63.

[6]   Logue, C. L. et al. (1989). Toxicity Screening in a Large, Municipal. Wastewater System. *J. Water Pollut. Control Fed.*, 61, 632.

[7]   Joann, S.; Richard, O. M.; Joseph, H. S.; Weber, A..S. and Michael, D. A. (1990). Activated Sludge. *Res. J. Water Pollut. Control Fed.*, 62 (4), 398-406.

[8] Eckenfelder, W. W. Jr. (1989). *Proceedings of The 43rd Ind. Waste Conf., Purdue Univ., West Lafayette, Ind.,* 1, 73-78.

[9] Patoczka, J. et al. (1989). *Proc. 43rd Ind. Waste Conf., Purdue Univ., West Lafayette, Ind.,* 51, 256-265.

[10] Watkin, A. T. and Jr. Eckenfielder, W. W. (1989). *Water Sci. Technol. (G.B.),* 21, 593.

[11] Chong, N. M.; Luong, M. L. and Hwu, C. S. (2012). Biogenic substrate benefits activated sludge in acclimation to a xenobiotic. *Bioresour. Technol.,* 104, 181-186.

[12] Harrison, E. M. and Barker, J. F. (1987). Sorption and enhanced biodegradation of trace organics in a groundwater reclamation Scheme-Gloucester site, Ottawa, Canada. *J. Contam. Hydrol.,* 1, 349-373.

[13] Shimp, R. J. and Pfeander, F. K. (1985). Influence of easily degradable naturally occurring carbon substrates on biodegradation of monosubstituted phenols by aquatic bacteria. *Appl. Environ. Microbiol.,* 49, 394-401.

[14] Moorman, T. B.; Cowan, J. K.; Arthur, E. L. and Coats, J. R. (2001). Organic amendments to enhance herbicide biodegradation in contaminated soils. *Biol. Fertil. Soils,* 33, 541-545.

[15] Nelson, E. C.; Walter, M. V.; Bossert, I. D. and Martin, D. G. (1996). Enhancing biodegradation of petroleum hydrocarbons with Guanidinium fatty acids. *Environ. Sci. Technol.,* 30, 2406-2411.

[16] Lee, K. (2003). Effect of additional carbon source on naphthalene biodegradation by Pseudomonas putida G7. *J. Hazard. Mater.,* 105 (1-3), 157-167.

[17] Swindoll, C. M.; Aelion, C. M. and Pfaender, F. K. (1988). Influence of inorganic and organic nutrients on aerobic biodegradation and on the adaptation response of subsurface microbial communities. *Appl. Environ. Microbiol.,* 54, 212-217.

[18] Corseuil, H. X.; Hunt, C. S.; Ferreira, R. D. S. and Alvarez, P. J. J. (1998). The influence of the gasoline oxygenate ethanol on aerobic and anaerobic BTX biodegradation. *Water Res.,* 32, 2065-2072.

[19] Lewis, D. L.; Kollig, H. P. and Hodson, R. E. (1986). Nutrient limitation and adaptation of microbial populations to chemical transformations. *Appl. Environ. Microbiol.*, 51, 598-603.

[20] Hu, Z.; Ferrainab, R. A.; Ericsonb, J. F. and Smetsa, B. F. (2005). Effect of long-term exposure, biogenic substrate presence, and electron acceptor conditions on the biodegradation of multiple substituted benzoates phenolates. *Water Res.*, 39, 3501-3510.

[21] Chong, N. M. (2009). Modeling the acclimation of activated sludge to a xenobiotic. *Bioresour. Technol.*, 100 (23), 5750-5756.

INDEX

F

G

H